Transportation Planning and the Future

Transportation Planning and the Future

PETER NIJKAMP, SYTZE A. RIENSTRA AND JAAP M. VLEUGEL
Vrije Universiteit, Amsterdam

JOHN WILEY & SONS
Chichester · New York · Weinheim · Brisbane · Singapore · Toronto

Baffins Lane, Chichester,
West Sussex PO19 1UD, England

National 01243 779777
International (+44) 1243 779777
e-mail (for orders and customer service enquiries): cs-books@wiley.co.uk
Visit our Home Page on http://www.wiley.co.uk
or http://www.wiley.com

Other Wiley Editorial Offices

John Wiley & Sons, Inc., 605 Third Avenue,
New York, NY 10158-0012, USA

WILEY-VCH Verlag GmbH, Pappelallee 3,
D-69469 Weinheim, Germany

Jacaranda Wiley Ltd, 33 Park Road, Milton,
Queensland 4064, Australia

John Wiley & Sons (Asia) Pte Ltd, 2 Clementi Loop #02-01,
Jin Xing Distripark, Singapore 129809

John Wiley & Sons (Canada) Ltd, 22 Worcester Road,
Rexdale, Ontario M9W 1L1, Canada

Library of Congress Cataloging-in-Publication Data

Nijkamp, Peter.
Transportation planning and the future / Peter Nijkamp, Sytze A. Rienstra, and Japp M. Vleugel.
p. cm.
Includes bibliographical references and index.
ISBN 0-471-97408-0
1. Transportation—Environmental aspects—Europe.
2. Transportation and state—Europe. 3. Environmental protection—Europe. 4. Sustainable development—Europe. I. Rienstra, Sytze A. II. Vleugel, Japp M. III. Title.
HE242.N54 1997
388'.068–dc21 97–34188
CIP

British Library Cataloguing in Publication Data

A catalogue record for this book is available from the British Library

ISBN 0-471-97408-0

Produced from camera-ready-copy supplied by the authors
Printed and bound in Great Britain by Biddles Ltd, Guildford and King's Lynn
This book is printed on acid-free paper responsibly manufactured from sustainable forestation, for which at least two trees are planted for each one used

Contents

PREFACE

Cities, regions and countries all over the world are witnessing a process of unprecedented change as a result of both integrative and disintegrative forces in a modern network society. Transport and communication play a strategic structuring role as the backbone of such developments. This central position of transport and communication also generates many intriguing research and policy questions. Is the transport sector able to fulfil its strategic economic role and is it able to overcome all institutional, financial and socio-political impediments? Is the drive towards more speed and accessibility a benefit for all regions and socio-economic groups or does it lead to new geographical discrimination and segmentation? Is the rapid increase in mobility of people and in flows of commodities compatible with environmental sustainability and conservation, or would the modern information society, through advanced telematics and electronic highways, offer a more appropriate solution? It is thus evident that modern transport is at the crossroads of strong, but also conflicting, forces: strategic economic growth, geographical accessibility and quality of life. These issues will certainly become more visible and pronounced at the beginning of the next century.

Although it has to be recognized that physical movement and spatial interaction are a *sine qua non* for a network economy, it cannot be denied that the benefits of a mobile and interactive society also cause many negative side-effects. In recent years, transport at all geographical levels has continued to increase congestion and traffic hazards, to produce disturbing levels of traffic noise, to use non-renewable energy sources and to emit greenhouse gases. Extrapolation of current trends towards the future would not only lead inevitably towards critical bottlenecks and serious environmental decay, but also to increasing disparities in the accessibility of most regions in our world. In addition, knowledge about the complexity of transport problems is still limited which further restricts efficient proactive policy making. Social and political approaches need to be recognized prior to, rather than after, the formulation of policy. There is still a shortage of clear insights into the ways actors in transport respond to alternative policy measures. These developments underline the urgency of a new strategic approach in future research on transport.

The present book addresses the above-mentioned issues by focusing in particular on the limits to mobility caused by the social costs of modern transport, not only locally but also globally. This is done from the long-term perspective of

trying to reconcile a further development of our mobile society with environmental constraints. The study thus addresses the problem of the compatibility of future transport systems with sustainability targets, and aims to map out new options by applying the so-called Spider methodology. The time horizon taken here is a long one (2030) and hence we had to resort to scenario experiments based on expert opinion. The background is the European situation, with particular reference to the Dutch case, but we believe that this situation is fairly representative of transport and mobility issues in most developed countries. This book contains a wealth of information from many sources and may be of interest to a varied readership: policy-makers and policy-making bodies, non-technical policy analysts and planners, the transport industry and regional and transportation scientists is the outcome of a broad research project on 'A Comparative Analysis of Options for Sustainable Transport in the 21st Century', undertaken in the framework of the Dutch National Research Programme on 'Global Air Pollution and Climate Change'. The scientific input by Richard Smokers (ECN), Johan Visser (OTB), Marina van Geenhuizen (TUD), and Ton Rooijers and Edith Van der Heijden (VSC) into the preparation of this study has to be explicitly recognized. Also the valuable contribution of members of the Dutch *Colloquium Vervoersplanologisch Speurwerk* has to be mentioned here. Furthermore, we wish to thank the Network
on European Communications and Transport Activity Research (NECTAR) for its support in evaluating our scenario results. The editorial assistance of Patricia Ellman has to be gratefully acknowledged as well.

Amsterdam, December 1997

Peter Nijkamp
Sytze Rienstra
Jaap Vleugel

Part A

SETTING THE SCENE

1 TRANSPORTATION IN A CONFLICTING FORCE FIELD

1.1 GENERAL BACKGROUND

The rapid rise in economic activity in the second part of the twentieth century has not only significantly contributed to the current unprecedented level of human welfare needs, but has at the same time caused a serious threat in the form of high social costs emerging from environmental externalities. Such external costs can be observed at present in all societies, in all regions and at all geographical levels, from local to global (Verhoef, 1996a). There is a worldwide concern for the quality of the environment ranging from local noise nuisance to global climate change or destruction of the ozone layer. Also the natural environment is facing a situation of structural decay reflected *inter alia* in loss of biodiversity or vulnerable ecosystems. Such ecological impacts not only affect the welfare position of the present generation, but also that of future generations. Consequently, the idea of environmental sustainability is at stake here, a problem mainly caused by the paradoxical situation that rational decisions and actions of individuals may lead to socially adverse impacts (i.e. the 'social dilemma' or 'social trap'). In general, it is a heroic and almost unviable undertaking to develop consistent and acceptable individual decision which would guarantee a balanced intergenerationally sustainable outcome (Whitelegg, 1993). The current emergence of a socially intolerable - unsustainable - outcome of normal human activities seems to oppose Mandeville's principle that 'private vices are public virtues'. Governments have also become part of the problem. Apparently, a blend of government and market failure is an intrinsic feature of modern democratic decision making, so clearly demonstrated by the NIMBY (not in my backyard) syndrome.

Policy instruments and measures for coping with environmental decay are certainly not unambiguous or undisputed, as their effectiveness and acceptance depend on the type of pollutant, the source type, the number of receptor areas, knowledge on cost functions and market imperfections (see Klaassen, 1995; Tietenberg, 1992, and Rietveld, 1997). Especially if all members of society contribute to environmental externalities and are also at the same time victims, a complicated environmental policy trade-off emerges.

A glaring example of the above-mentioned 'tyranny of small decisions' (Kahn, 1966) can be found in the transportation sector, where numerous individual mobility decisions create an overall problem in terms of congestion, environmental pollution, traffic safety and landscape destruction (see also Gärling, 1993). In the framework of the worldwide sustainability debate, much attention is given nowadays to the CO_2 emissions of the transport sector, but it is clear that the social costs of transport on both a local and a global scale cover a much wider array of externalities.

The industrialized world appears to have the highest transport mobility rates. In particular, its economic heartlands have to digest unprecedented volumes of traffic. Clearly, transport is a necessary part of economic development but it is also responsible for a wide range of negative (i.e. unpaid) externalities in the form of congestion, traffic accidents, environmental pollution, landscape destruction, unand solid waste. According to recent estimates, nowadays the social costs of transport amount to some 3 to 5 per cent of GDP in most OECD countries (see e.g. Barde and Button, 1990; Himanen *et al.*, 1993; Verhoef, 1994, 1996b) and this figure is gradually rising. Consequently, transport activities do create an unsustainable development.

A wide range of policy measures have been proposed to cope with the high social costs of mobility, e.g. information campaigns, user charges, standards on emissions, constraints on mobility, new forms of land use and physical planning, and new transport technologies (Garett and Wachs, 1996). Experiences from different countries and cities all over the world have shown that there is no single unambiguous and effective remedy. Instead, there is a range of possible solution strategies, none of them providing a final answer. Some authors (e.g. Jones, 1991) suggest the use of a package approach to transport sustainability policies in order to increase efficiency and effectiveness.

It has to be recognized that the awareness of the environmental consequences of the transport sector has risen dramatically in recent years, but the political system has largely been paralysed in taking effective measures (Ostrom, 1990; Rienstra and Nijkamp, 1997a). Apparently, transport is playing a central role in a force field of multiple conflicting actions and views, so that consensus on a sustainable road towards the future is hard to find (Nijkamp, 1994). There are conflicting norms and values, there are individual and collective resource and situational constraints, there are strict economic scarcity phenomena and a myriad of other factors which form impediments to a smoothly functioning network society. The case of governance failures especially is noteworthy here, as policy attempts to curb mobility have largely failed due to bureaucratic decision-making procedures, lengthy regulatory and participatory procedures, and insufficient knowledge of the behavioural responses of individuals. Nevertheless, there is a need for a regulation of the transport market which would reduce the present unacceptably high social costs. Public policy is weak in this setting, as a frame of reference substantiated by scientific knowledge is either missing or not politically accepted.

The present study aims to fill this gap in strategic thinking by offering a methodology for scenario analysis in which expert knowledge is integrated with policy ideas on the future of our spatial and transportation systems with a view to achieving a more environmentally friendly mobility pattern. Thus the analysis in this book will mainly be concerned with a search for a sustainable transport system, where the necessary flows of people or commodities and the underlying infrastructure provisions are in agreement with relevant and acceptable (socio-) economic goals within the limits set by the 'environmental utilization space', in such a way and to such an extent that the interests of future generations are not jeopardized (Vleugel, 1995). This general description, which is largely in agreement with the Brundtland definition (WCED, 1987), does not provide a 'hard' measurable criterion, but rather an orientation point for the analysis undertaken in the present study.

1.2 DEMARCATION OF THE STUDY

Transport is largely a derived demand. Consequently, it is almost impossible to isolate the transport sector and its related mobility pattern from all other human and industrial activities. Thus any attempt at investigating the environmental impacts of transportation behaviour and policies is necessarily a multifaceted undertaking. Nevertheless, the need for a broad spectrum approach runs the risk of lack of focus and hence of spurious conclusions. Therefore in our study on strategies for sustainable transport systems various limitations have to be built in.

1.2.1 Geographical scale

It will be necessary to demarcate the geographical coverage of the study, without losing a sense of generality. A choice is made in favour of realistic options and developments by taking Western Europe as a geographical frame of reference. This is not a major restriction. The patterns observed in Western Europe are fairly typical of most industrialized countries. We need, however, a base line case for constructing future scenarios, and from this perspective the European situation offers interesting and practical illustrations. This area is chosen because it is largely homogeneous in terms of transport demand and supply compared to other regions, such as Africa, Asia or Latin America. The most important reason for restricting ourselves to the Western world is its present - dominant - share in world energy use (hence in global warming): the OECD countries account for about 66 per cent of the world's energy consumption in transport; strikingly, this share has risen between 1971 and 1990 (Grübler, 1994; see also Chapter 2). It seems plausible that there are still many possibilities in the Western world to reduce its contribution significantly; otherwise, a sustainable transport system at a worldwide scale is a utopia. A higher spatial scale, notably the global scale, would complicate matters more than it would serve the purpose of this study.

The empirical analysis is mainly based on Dutch field work, but the Netherlands is typical of many Western European countries. Nor does this restriction mean that developments and experiences from other countries are not considered (e.g. the Japanese Shinkansen or tunnel construction or the US air transport organization). A logical limitation of this choice is that one may eventually have to conclude that some of the transport systems or scenarios discussed may not offer all or even sufficient answers for specific transport issues in other countries. On the other hand, a smaller scale analysis (e.g. regional or national level) would mean that certain highly interesting modes of transport - from an environmental or transportation point of view - would have to be left out (e.g. air transport and high-speed trains - HST). However, more local environmental issues are not neglected in this study; they will be considered whenever necessary. In this respect it should be recognized that urban areas in particular are facing serious environmental problems due to congestion and local externalities.

1.2.2 Time horizon

A prospective and exploratory analysis becomes less credible and more risky when the time period considered lies too far away. This calls for reducing the relevant period. We have therefore chosen by and large the year 2030 as the limit for our analysis. The time span is long enough to allow for rather drastic changes, and short enough to connect current trends to new futures. This does not mean that foreseeable developments with a shorter lead time are left out, but we assume that various short-term technical - and other - barriers will be overcome in due time, and consequently do not halt the realization of developments in the not too distant future. Our analysis of the future is preceded by a retrospective view on the period 1970-1990 (Section 2.4).

1.2.3 Passenger versus freight transport

Passenger and freight transport differ in many respects, though they also have various aspects in common and cannot be entirely separated. For the purpose of focus and manageability, we have chosen to address mainly technical and non-technical options and scenarios for passenger transport. This restriction looks at first sight rather constraining as freight transport also leads to various environmental issues which should not be overlooked, especially since freight transport is growing apace. Studying freight transport would, however, require an investigation of far more options and transport actors, since the problems associated with shipping goods are largely different. Despite these limitations and specificities, it is clear that many links do exist between passenger and freight transport. For instance, in many cases both categories use the same infrastructure, logistics and propulsion systems. What we are really interested in is a more general picture of the viability of new transport systems technology. If these innovations work for passenger transport, they may to some extent also work for freight transport. Further analysis may reveal whether this is the case or not.

In the course of this study many other specific assumptions and technological approaches will be introduced, but these will be discussed in the relevant sections.

1.3 SCOPE AND PLAN OF THE BOOK

Transport policy has become one of the strategic tools to cope with environmental decay. It is noteworthy that, in general, transport policy follows a conventional strategic cycle, starting with (1) incremental improvement measures; (2) organizational managerial changes; (3) extension investments in infrastructure; (4) technological improvements; (5) financial charges to users, followed by (6) more radical technological changes (e.g. high speed trains, computerized transport telematics, etc.) (see also Figure 1.1).

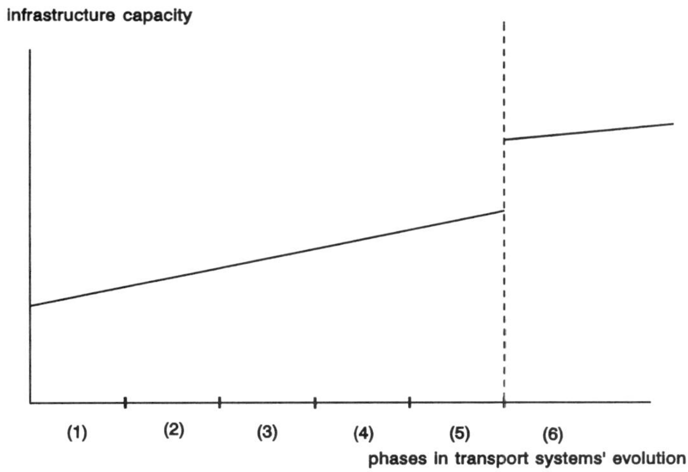

Figure 1.1. Life cycle of transport systems' capacity

Thus it seems that technological jumps are normally at the end of the life cycle of the existing transport product, as they are a response to insoluble problems in the prevailing mobility systems. A major analytical and policy problem is, of course, when the new transport technology will come into being and whether it will offer in time a proper response to all emerging frictions in transport systems. The prediction of such changes is fraught with many uncertainties, even though logistic growth curve analysis may *ex post*, provide some lessons (see Grübler and Nakićenović, 1993). In this context, future scenarios may play an important role in scanning the uncertainties, in designing new opportunities and in identifying possible bottlenecks which are likely to influence the achievement of sustainable transport systems for the next generation.

This book offers a strategic analysis of key factors that may act as impediments to a sustainable transport system. A change in such factors may allow us to identify the driving forces that may lead to a more environmentally-friendly mobility pattern. The main background conditions taken into consideration in this study are spatial, institutional, (socio-)economic and socio-psychological factors. These factors are assumed to be of decisive importance for mobility patterns in our modern network economy. Technological factors are of course extremely important as well, but are mainly seen as responses to the needs of a mobile society driven by the above-mentioned force field. This also means that technology has an ambiguous role in the context of a sustainable transport system. It may be a benefactor in that various externalities of the transport sector may be alleviated, but it may also be the source of much harm in the sense that a technological response to meet the demands of *homo mobilis* may worsen environmental quality conditions.

This study will make an attempt to depict the above force field, not only conceptually, but also empirically by designing and assessing various scenarios of future transport systems, based on expert opinion. Thus an important methodological contribution of this study is the development and evaluation of expert-based scenarios in which 'future realities' and 'future wishes' are systematically confronted with one another. Another main feature of this study is that it offers a wealth of empirical data and views on major trends in current transport systems, so that it also may act as a source of orientation for planners and policy makers.

In line with the above observations, the organization of this study is as follows. The book is subdivided into four parts. Part A presents an overview of current trends in transport and its environmental impacts. In Part B the discussion is focused on technical innovations which are already developed or currently under development, as well as their success and failure factors. Next, Part C maps out reference and expert scenarios for the future of transport systems by applying the newly developed 'Spider model', which depicts both expected and desired trends in the above-mentioned background factors. Finally, strategic policy conclusions are drawn in the last chapter.

2 TRANSPORT AND THE ENVIRONMENT

2.1 INTRODUCTION

Over many decades transport and traffic have grown at a steady pace, and this trend is likely to continue because of a combination of various demand and supply factors (Grübler, 1993). In the post-war period public and political awareness of the harmful negative impacts of transport and traffic on the natural environment and living conditions of many people was not strongly developed. The negative externalities of transport were recognized when land became more scarce, when most natural environments were cultivated, when global change phenomena started to emerge, when local living conditions increasingly deteriorated through the effects of traffic, and - even more importantly - when the negative impacts of environmental externalities on the economic performance of societies became only too apparent. For example, it is now estimated, that the non-internalized social costs caused by transport may account for some 5 per cent of GDP in the European Union (*Monitor*, 1994; Possum, 1997). As a consequence, the environment has become a highly valuable asset instead of a dispensable one, in both the minds of citizens and in official policy documents.

In this chapter a global description is presented of the environmental impacts of transport in order to analyse in which way and to what extent solutions are needed and changes are necessary. First, local/regional problems are discussed. This will be done briefly, as this book largely focuses on global environmental problems. Next, a short general description is given of the emission of greenhouse gases, followed by a description of those caused by transport. In this book we focus mainly on the emissions of greenhouse gases - especially those of CO_2 - as the main sustainability indicator, although we will also take various other externalities into account.

2.2 THE LOCAL ENVIRONMENTAL IMPACTS OF TRANSPORT

The negative environmental impact of transport covers a broad spectrum of concerns at different levels. The most important ones at a local/regional level are discussed in this section. Since the scope of this book is mainly geared to global issues, this description concentrates on the essentials (see also ECMT/OECD, 1995; Gwilliam and Geerlings, 1994; OECD, 1988b; Short, 1992).

2.2.1 Local air quality

Transport contributes not only to the deterioration of the global atmosphere, but also leads to local air pollution because of its emissions of CO, HCs (hydrocarbons - aldehydes, benzene, etc.), particulates (lead, asbestos, etc.), ozone (photochemical smog) and NO_x. Local air quality is poor in many metropolitan areas and is a source of serious concern (see Nijkamp *et al.*, 1996a). In the Western world, we find the most negative examples in heavily motorized countries like the United States (e.g. Los Angeles). But also many European cities suffer from severe air pollution, e.g. many Italian cities, or Athens in Greece. Increasingly, local air pollution also becomes a problem in the developing world (e.g. Mexico City, Bangkok, Djakarta). In many cases the causes are the same: an unfavourable combination of high population density, a high-density road network, and a massive - and still growing - car use, combined with a low-quality (or overcrowded) public transportation network. Specific geographical conditions make up for the rest (see also ECMT/OECD, 1995). Research in the Netherlands has shown that on congested roads in inner cities emission levels are twice as high as those found for the Netherlands on average (Langeweg, 1988).

The impact of local air pollution is not only felt by road users, such as motorists, cyclists, playing children and pedestrians, but also by those urban residents living alongside such roads. All these people face roughly the same problems, most notably a lower productivity as well as health diseases related to air pollution (e.g. asthma, lung diseases, cancers).

2.2.2 Acidification

Many human activities contribute to the acidification of the environment; most notably agriculture (NH_3 emissions), industry, transport and heating of dwellings and buildings. Transport emits nitrogen oxides (NO_x), sulphur dioxides (SO_2) and non-methane hydrocarbons (NMHCs). Environmental impacts are *inter alia* reflected in the large-scale deterioration or decay of forests all over Europe, and also at a worldwide scale.

2.2.3 Solid waste

The production and use of vehicles generates large amounts of solid waste (often of a dangerous nature) which have to be disposed of, either in car dumps or by shredding and recycling. The millions of cars - still rapidly increasing in numbers - cause an enormous solid waste problem for our societies.

2.2.4 Disturbance, interference and intrusion

Transport leads to various forms of nuisance in both the man-made and the natural environment, as a result of the use of vehicles (noise, bad smell) and infrastructure (e.g. visual intrusion). The face of modern cities is largely determined by the transportation sector.

2.2.5 Exhaustion of resources

Another environmental impact is the waste or exhaustion of the earth's resources for both vehicles (e.g. fossil fuels, metal, rubber and oil) and the infrastructure (e.g. sand, gravel, tar, oil, iron).

2.2.6 Segmentation of landscapes

Finally, there is the spatial segmentation and mutilation of the (natural) environment, for both nodal infrastructure (parking places, stations, airports, harbours), line infrastructure (corridors) and network infrastructure (roads, railways, etc.).

The disappearance and extinction of many animal species and plants may partly be related - directly or indirectly - to the growth of transport and its infrastructure. Many diseases, such as lung cancer, asthma, or sleep disturbances, may in some way or another also be linked to the effects of modern transport. If this impact continues, many of these externalities are also likely to restrict the living conditions of future generations and non-human species. Even though such externalities may manifest themselves at local or regional scales, the sum total of all such impacts may have a worldwide influence on the quality of ecosystems and the human environment.

2.3 CLIMATE CHANGE

2.3.1 Introduction

Climate change covers both the greenhouse effect and the depletion of the tropospheric ozone layer (see also Houghton *et al.*, 1990). In this book we will mainly focus on global air pollution, since this may be the hardest externality to alleviate. Global air pollution is caused by emissions of CO_2, NMHC, N_2O, CFCs and tropospheric ozone. Even though CO_2 is strictly speaking not a pollutant (as it is a natural component of any life-support system), it is generally expected to have a major impact on climate change.

The depletion of the ozone layer is partly caused by human emissions of aerosols (CFCs) and nitrogen oxides (NO_x). Recently, there has also been empirical evidence that volcanic activity, especially the output of sulphurous particulates, may significantly contribute to the depletion of the ozone layer. This depletion is an issue which is at least as serious as the greenhouse effect, as it may *inter alia* lead to an increase of many skin-related diseases such as cancers and eye diseases. It may also reduce the yield of many essential vegetables and crops, thereby complicating the food situation and thus the health component of socio-economic conditions in many countries.

The greenhouse effect is a complex problem, which is only partly understood at the moment. On the one hand, many human activities, including transport, cause emissions of 'greenhouse gases', most notably CO_2, methane (CH_4), NMHCs and CFCs (Martin and Michaelis, 1992). As well as these gases, the output of particulates by volcanos may also contribute to the greenhouse effect.

On the other hand, there is the balancing absorption capacity of the environment (or 'sink'; most notably by the oceans).

It has been argued by many scientists that the output of greenhouse gases exceeds the absorption capacity of the environment. Once this imbalance reaches a critical level, the average temperature of the earth is likely to rise. This may lead to many environmental and land-use problems in various countries, for example, a general rise in the levels of oceans and seas, which will cause more floods (see Van den Bergh and Nijkamp, 1996). Besides, it is likely that wind and rain patterns will change, probably bringing more extreme weather conditions in their wake. Another serious problem is the loss of agricultural areas due to salinization, etc. The greenhouse effect still has many unknown aspects, but in general it is accepted that its impacts may be quite drastic for many aspects of human life.

The use of all these carbon-based energy reserves in combination with present technology may therefore be detrimental to both the natural and the built environment. Consequently, it seems a plausible policy strategy to adapt a 'no regrets' principle for policy-making regarding the greenhouse issue, even though the precise impacts are not yet known. In such a strategy, efficient environmental and resource policies are identified and preparations for their introduction are made. However, policy measures are only introduced when they also serve other objectives; governments can then respond efficiently when more rigorous policies appear to be necessary in due course. The most important greenhouse gases and their impacts are discussed in the next section.

2.3.2 The main greenhouse gases

The main gases causing climatic change have been summarized in Table 2.1, together with their global warming potentials (GWP). The GWP indicates how 'strong' the influence of the gases is on the greenhouse effect. The total impact of course also depends on the quantity of the emissions.

It appears that many gases are quite stable in chemical terms, which means that they dissolve very slowly in the atmosphere. In other words, most of what we emit today will still be there in 20, 100 or even 500 years' time. If we do not stop emitting these substances, they build up in the atmosphere and will have detrimental effects, both in terms of global warming, and in terms of reducing the ozone layer (CFCs). From this perspective, the plans for the reduction of human CO_2 emissions and the worldwide consensus on the ban on CFCs are hopeful signs. But the figures also show that the substitutes for CFCs (the HCFCs) are far from harmless, at least in terms of GWP. A comparison of the 1992 figures with the 1990 figures tells us that more recent knowledge about the physical-chemical nature of processes in the atmosphere has for most substances not led to drastic changes in the estimated climatic impact of these substances. Some further evidence can also be found in the 1996 IPCC report.

We now present a concise overview of the main greenhouse gases; however, we will not discuss methane (CH_4): for this, the interested reader may refer to Heilig (1994).

Table 2.1. Global Warming Potentials (GWP) of selected greenhouse gases, mass to mass base (GWPs are dimensionless)

Greenhouse gases	(IPCC[1] 1990) Time horizon (years)			(IPCC 1992) Time horizon (years)		
	20	100	500	20	100	500
CO_2	1	1	1	1	1	1
CH_4 direct	26	6	2	35	11	4
CH_4 chem.	37	15	7	NQ[2]	NQ	NQ
N_2O	270	290	190	260	270	270
CO	7	3	2	NQ	NQ	NQ
CFC-12	7100	7300	4500	7100	7100	4100
HCFC-22	4100	1500	510	4200	1600	540
HFC-134a	3200	1200	420	3100	1200	400
HFC-152a	510	140	47	530	150	49
NO_x	30	8	3	NQ	NQ	N.Q.

Notes: 1) IPCC = International Panel on Climate Change
2) NQ = Not quantified

Sources: Houghton *et al.* (1990; 1992); Ybema and Okken (1993)

CO_2

This is an absolutely crucial gas for the living processes of the earth. This explains why there is a long CO_2 chain which links the atmosphere, the oceans and land surfaces and all the species living in them. Only part of this CO_2 is 'free' CO_2, which means that most of this gas is stored in 'sinks'. Both oceans and forests are able to store large amounts of it. As long as this chain is closed, the amount of free CO_2 in the atmosphere is stable; there is a balance between emissions and absorption.

From historic data of the last century we know that the amount of free CO_2 in the atmosphere has grown considerably. This correlates with the steep growth in the earth's human population, which has led to the increasing use of fossil fuels and the reduction of CO_2 sinks such as forests. In other words, human activity has led to an imbalance in the emission-absorption cycle of the earth. At present, there is a stock of some 760 gigaton's (Gt) of carbon in the atmosphere, which is growing by about 6 Gt per year, mainly because of the use of fossil energy. From these 760 Gt, 200 Gt may originate from burning fossil fuels. All economically recoverable energy reserves amount to 540 Gt of carbon, while estimated resources measure 3000 Gt (identified resources, with uncertain recoverability prospects) and there are additional occurrences amounting to 5000 Gt (with assumed quantities having largely speculative technical and economic potential). It may be clear that burning these stocks would have an enormous impact on the greenhouse effect (Messner and Nakićenović, 1993).

An indication of the scale of reduction needed in the emissions of CO_2 can be found in the Toronto Declaration, which states that CO_2 emissions should be reduced by 50 per cent at a global level in the year 2025 (WMO, 1988). For Western countries in particular this would probably imply a reduction of 80 to 90 per cent to compensate for the expected increase of CO_2 emissions by developing countries. These targets have been agreed by many Western countries as a commitment for future emission reductions.

Notwithstanding the fact that the production of CO_2 is caused by many factors which are at present hard to control (e.g. population growth), one thing that may be done in the short and medium term is to save energy. This may be realized by improving end-user energy efficiency in the Western world, which uses a disproportionate amount of the earth's energy resources. The current global conversion rate from primary to final energy is 74 per cent, whereas the conversion rate of primary into useful energy is on average only 34 per cent (Messner and Nakićenović, 1993).

Putting these CO_2 policies into practice means a structural socio-economic and technological change. But such a change has a series of limitations and drawbacks, which may act as barriers for realizing the policy choices (see Chapter 4).

It is noteworthy from the information in Table 2.1 that CO_2 is not a very 'strong' greenhouse gas. Even though this gas is emitted in very large quantities, it is estimated to account for about 50 per cent of the greenhouse effect.

CFCs, HCFCs and HFCs

Another group of gases which contribute significantly to the depletion of the ozone layer are chlorofluorocarbons (CFCs), which are used in manufacturing spray foams and vehicle climate systems (air-conditioning) (Martin and Michaelis, 1992).

The CFCs and some of the HCFCs and hydrofluorocarbons (HFCs) are strong greenhouse gases, since they have a large GWP (see Table 2.1). In an indirect way, however, through the reaction with ozone in the upper troposphere, CFCs may also have a cooling effect on the atmosphere. The indirect cooling effect of HCFCs is much smaller as their potential to destruct ozone is much less. The HFCs have no cooling effect. The magnitude of the indirect cooling effect is currently very uncertain. For CFCs it may possibly balance its global warming potential, but HCFCs and HFCs most certainly have a positive net GWP. The combined contribution to the greenhouse effect of CFCs, HCFCs and HFCs is estimated to be some 17 per cent (Houghton *et al.*, 1992).

The awareness that the depletion of the ozone layer may reach dramatic proportions has recently led to a complete ban on the production of 'hard' CFCs in 1997. As these are used in vehicle air-conditioning and cooling systems, replacement by less harmful ones has become necessary. Leakage of these substances is a serious problem, so anti-leakage systems should be developed. The use of climate systems in vehicles as such may become a policy issue, as they also use a lot of energy. In this study the CFC issue is not further elaborated, since it is of lesser importance for transport.

Tropospheric ozone

While in the stratosphere ozone is depleted by CFCs, in the lower troposphere ozone is produced from photochemical reactions (smog). Stratospheric ozone filters UV-light and is therefore essential for the health of all living species on earth. In contrast, tropospheric ozone is a direct health hazard, and also contributes to the greenhouse effect. This contribution may be as high as 8 per cent. Through this mechanism the NO_x and hydrocarbon (HC) compounds in exhaust emissions add to global warming.

2.4 TRANSPORT AND THE EMISSION OF GREENHOUSE GASES

Transport causes a considerable part of the greenhouse gas emissions (see also Hughes, 1993). Its share of NO_x emissions in the EU is 58 per cent; the share of CO_2 emissions is 23 per cent (*Monitor*, 1994).

Figure 2.1 shows trends in CO_2 emissions of major sectors in the European OECD countries in recent years. It may be concluded that the emission growth rates caused by transport are much higher than those of other sectors and that transport has contributed significantly to the overall growth in CO_2 emissions in the years 1970-1990. This growth may be caused by several factors such as the increase in car ownership, rising mobility rates, etc. (see Nijkamp, 1994 and Chapter 3). As a result the energy use by transport has increased considerably. It should be added, however, that the reduction in CO_2 rates in the industrial sector may not only be ascribed to energy and environmental policies in this sector, but also to the general decline in the industrial manufacturing sector as a whole (a phenomenon characterized often as the 'post-industrial society').

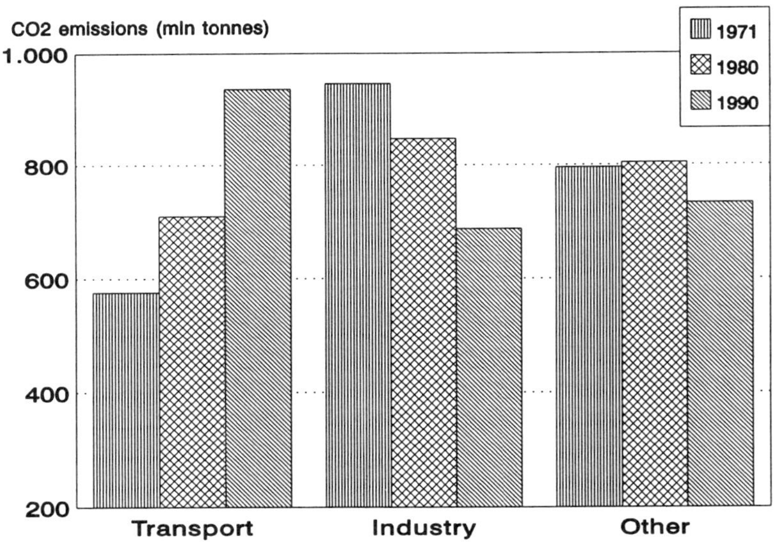

Figure 2.1. CO_2 emissions in European OECD countries by sector
Source: OECD, 1993a

In Figure 2.2 some interesting trends in air pollution by 'mobile sources' are presented for Western Germany. Trends in other Western countries will probably not differ significantly from this picture, although most countries may be lagging behind in the introduction of 'unleaded' fuels and the catalytic converter.

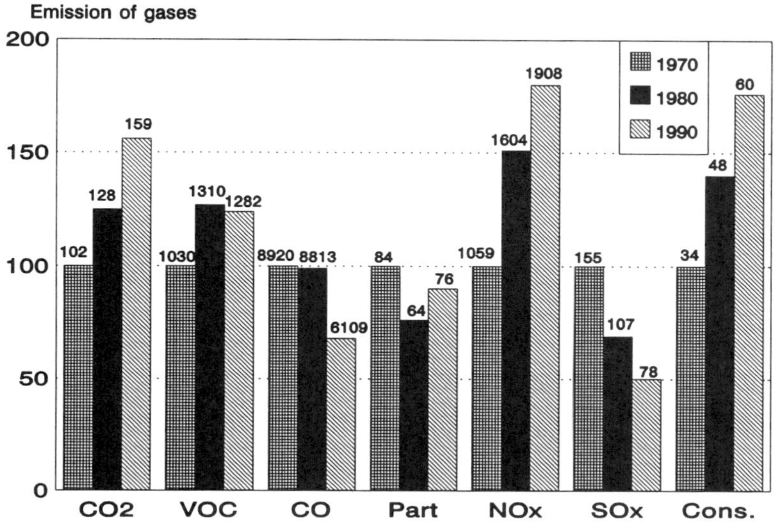

Figure 2.2. Emissions of gases and energy consumption by mobile sources in Western Germany (index: 1970 = 100)

Notes: 1) The 1990 figure of CO_2 refers to 1991
2) CO_2 is measured in mln tons; the other gases in kilotons; energy consumption in mln tons
3) Cons = final energy consumption by transport

Source: Own calculations based on OECD (1993a)

Table 2.2 shows that road transport has the highest CO_2 pollution per passenger kilometre. As will be discussed shortly, the most polluting modes also account for the highest modal share and growth rates in the past few decades.

Table 2.2. Modal comparison of CO_2 emissions in g/pass.km (index)

Pass. train	Car	Tram/metro	Bus	Plane
71	201	100	159	176

Note: These figures should be used with care, as the uncertainty margin may be up to 25 per cent or more, depending on the assumptions regarding seat occupancy, technology used, vehicle use, etc.

Source: Information provided by the Netherlands Railways (NS)

Table 2.3 indicates the 1988 greenhouse impact of major modes in passenger transport in the European Union (EU). This table shows the dominant position of road transport in the contribution of transport to the greenhouse effect, due to its high modal share and relatively low fuel efficiency. It also shows the fact that aircraft contribute to the greenhouse effect mostly by their emissions of NO_x, rather than by their CO_2 emissions.

Table 2.3. The 1988 greenhouse impact of transport in the European Community

	CO_2	CO	NMHCs	CH_4	NO_x	N_2O
	In Mtonnes CO_2 equivalent[1]					
Diesel car	280.4	5.4	3.2	1.9	14.3	1.1
Gasoline car	432.2	41.4	27.9	60.2	18.4	28.8
Electric train[2]	17.9	0.0	0.0	1.4	0.2	0.4
Aircraft	85.3	1.0	0.5	0.1	131.0	2.7

Notes: 1) Model estimations (+/- 20 per cent)
2) Including emissions from power plants
Source: Martin and Michaelis (1992)

It may be concluded that in the road transport sector the emissions of greenhouse gases have grown faster than those in other sectors, while the most polluting modes are also the most dominant ones. Therefore, it is evident that changes are unavoidable if these emissions are to be reduced. In this book we will mainly concentrate on CO_2 emissions, as these are caused by all modes of transport (in contrast to, for example, NO_x which is largely generated by aircraft). From the figures presented above, it would already at the outset appear that a reduction in CO_2 emissions may be extremely difficult to achieve.

2.5 EVALUATION CRITERIA FOR THE FUTURE OF TRANSPORT

From various studies (see e.g. WCED, 1987; Enquete-Kommission, 1994) it has become clear that the current level and nature of human activities are to a considerable extent incompatible with the environmental needs of current and future generations. This means that, if society does not make a transition to environmentally more appropriate ways of consumption, production and transportation, not only is the environment at stake, but also the consumption and production prospects of future generations will be adversely affected. In this respect, one should also appreciate that their needs are likely to be much higher, because of worldwide population growth and a predicted upward shift in average wealth.

This leads to the conclusion that consumption and production processes - and most notably (but not exclusively) the embedded technology - have to change drastically in order to offer future generations at least the same opportunities as we enjoy today. The application of this 'strong sustainability criterion' (Opschoor, 1992; Klaassen and Opschoor, 1992) implies that human activities should aim at not reducing a given stock of environmental resources any further. In this context the notion of the 'pollution bubble' is relevant. This means that pollution in a given geographical area, say a country, should not exceed a general critical threshold level. Actors (e.g. sectors) may only emit into the environment a level of pollution which matches the natural absorption capacity for that specific type of pollution. As soon as some activity exceeds the agreed pollution levels, other activities in the same bubble have to reduce their pollution. In this way, it is guaranteed that the overall stock of environmental resources will remain within targeted levels. This compensation scheme may be extended to other sectors of the economy or other countries, for example. It may take various forms (e.g. lower consumption of resources, better extraction technologies, new resources). A key element is thus that the amount of pollution should be smaller than, or equal to, the absorption capacity (or 'environmental utilization space') of the environment (see also Opschoor and Weterings, 1994). In general, a shift of pollution between scale levels (local-regional-global) or types (soil-air-water) should be avoided, while irreversible effects should also be prevented. This idea leads - despite the problems of feasibility - at least to some measurable sustainability criteria. In this context, tradeable permits in the transport sector may also offer new interesting opportunities (see for more details Verhoef *et al.*, 1997; see also Section 2.6).

2.6 TOWARDS POSSIBLE SOLUTIONS IN TRANSPORT

From the above observations it may be clear that large-scale future changes are necessary for reducing the CO_2 emissions and other externalities. In general, two kinds of options are available to reduce the external impacts of transport (see also Enquete-Kommission, 1994; Gwilliam and Geerlings, 1994):

- **technological** options, more specifically both an improvement (technical, managerial) of existing vehicles and modes and the introduction of new (more sustainable) technologies;

- **behavioural** options, namely a modal shift towards more environmentally friendly (nowadays mostly collective) modes and a reduction of the total mobility level.

It is clear that technological options will meet the least resistance in society, because no significant changes in behaviour are necessary and the costs of 'conversion' are more or less hidden from the users. Given this situation it is attractive to stimulate R & D in order to induce favourable developments. It is questionable, however, how much reduction in environmental externalities is likely to take place as a result of such technological developments (see also Vleugel, 1995).

On the other hand, the behavioural options may generate fierce resistance, since in that case more or less structural changes in society, e.g. in lifestyle, leisure culture, social and psychological behaviour, are necessary, which will affect economic efficiency and lead to - possibly prohibitively - high transaction costs. The level of these transaction costs may be lowered, however, by using appropriate policy packages. Although there is a lot of experience with 'repressive' policy measures (higher user prices, traffic and parking regulations), this experience is not particularly positive; changes in user behaviour are either short-lived or even absent and some undesirable effects might also be induced. This calls for developing new policy tools. In recent years various authors (see, for instance, Goddard, 1997 and Verhoef *et al.*, 1997) suggest that tradeable permits may be used to change consumer and producer behaviour in an environmentally more friendly direction. Tradeable permits are said to combine the positive effects of both pricing instruments and regulatory instruments, but not - or only part of - their negative side-effects. So, it may be an efficient and cost-minimizing solution strategy. It is also interesting that the idea of tradeable permits may be combined with the idea of 'environmental utilization space', which was discussed at the end of Section 2.5.

As this study focuses primarily on technology, these behavioural options are not discussed in more detail. Part B contains an analysis of success and failure factors of technological options which have been developed or are now being developed in the transport sector. This analysis should give us an answer to the question of how far they may contribute to sustainability, bearing in mind that there is enough evidence to suggest that they may at best offer only a partial solution. This means that behavioural options should be stimulated with at least the same vigour.

First, however, the next chapter contains an overview of current trends in transport in order to provide a total picture of the scene in which transport operates today and in the near future.

3 MAJOR TRENDS IN EUROPEAN TRANSPORT

3.1 INTRODUCTION

In the second part of the 20th century a new human species has emerged, the *homo mobilis*, whose needs satisfaction is dependent on transport. Transport, a generic term referring to the movement of persons, goods and information in space, has become one of the key economic activities in modern societies. As mentioned before, it is a derived activity, which means that for most individuals it is a way to facilitate other human activities. As a result, transport assists our daily life in various ways, both in passenger transport (e.g. commuter traffic) as well as in goods transport.

Transport has become a prerequisite for the proper functioning of modern societies, but - as demonstrated in Chapter 2 - there are also various drawbacks, most notably its negative impact on the environment and on the quality of life.

In this chapter several trends are analysed that are currently emerging in the transport sector as well as in other fields which are of direct or indirect importance for transport. These fields are related to spatial, institutional, economic and socio-psychological aspects of mobility behaviour.

First, European environmental and transport policies are briefly discussed, followed by a general discussion of West European trends in the transport sector. Also some attention is given to the transport policy in distinct countries. Then trends in related fields are described. And finally, we present some empirical evidence concerning which transportation trends, behavioural developments and environmental consequences are seen as plausible by transportation experts.

3.2 THE EUROPEAN ENVIRONMENTAL AND TRANSPORT POLICY

3.2.1 Introduction

The common European transport policy has several aims (Commission of the EC, 1993a; Wenke, 1992; Hey, 1996). In the first place, the development of trans-European networks is being stimulated, which should also favour the development of backward or less favoured regions. Secondly, the transport markets have to be liberalized as much as possible; regulations should in principle be equal in every Member State and the markets have to be opened up for agents from every Member State. And thirdly, the aim is to achieve sustainable mobility (see also Nijkamp *et al.*, 1994), an objective which may be realized in various ways, e.g. by imposing market principles aimed at internalizing the external costs of transport.

The problem is that merging economic needs with ecological ones may be fraught with almost insurmountable difficulties. In this section attention is given to the common European policy regarding transport and infrastructure. It should be mentioned, however, that the policy is more of a stimulating and initiating nature, while - according to the subsidiarity principle - the final responsibility for implementation and enforcement still largely rests with the individual Member States of the EU (Duenk, 1991). Another main problem for a common European policy is the lack of funds, although in recent years several more sources of funds have become available.

3.2.2 The European transport policy

European regulation aims at reducing air pollution by road vehicles in various ways:

- by setting emission reduction targets per vehicle;
- by reducing traffic congestion;
- by reducing mobility growth.

The first of these measures has been successfully applied over recent decades in Europe. A reduction of NO_x, HC and CO emissions up to 80 per cent has been achieved in new cars since the 1980s by using catalytic converters and by increasing the engine efficiency. The latter measure has led to a reduction of fuel use and thereby lower CO_2 emissions per vehicle. Yet, because of the simultaneous rise in mobility the net energy consumption and emissions of CO_2 by transport have increased (see also Chapter 2).

At present, there is no common European regulation regarding exact targeted levels of CO_2 emissions; the official target of the EU is to achieve again the 1990 emission level of CO_2 (which has since then been rising) in the year 2000. This general target is not compulsory, however, so that separate targets have to be set and implemented by the national governments. The different targets of the governments of several countries are presented in Table 3.1.

It appears that all EU-members, except Ireland and Spain, have made the EU target an official policy goal, while several countries have an even more ambitious goal. Various countries have also set long-term targets for their CO_2 emissions or have adopted specific targets for the transport sector and introduced an active (transport) policy to achieve these goals (see also Salomon *et al.*, 1993). For example, the current Dutch transport policy target for CO_2 emission reduction in the period 1986-2010 is -/-10 per cent(Tweede Kamer, 1990b), Denmark aims at a stabilization in 2005 and a 25 per cent reduction by 2030 (ECMT/OECD,

1995), while Germany intends to reduce these emissions even more speedily - by 25 per cent in 2005 (Wissman, 1994). Switzerland plans to reduce its emissions mainly by stimulating public transport (Andersen, 1993). Countries like Spain, on the other hand, have not formulated strict policies to reduce the external effects of transport; in fact the policy goal is to stimulate mobility in order to achieve economic growth (Cedrún and de Rus, 1994).

Most governments try to achieve these goals by a mix of several policies (see also Section 2.6). One of them is reducing traffic congestion. This may help to reduce air pollution and energy use. This aim may also be achieved by means of traffic management schemes or by constructing more infrastructure for lower emission modes (e.g. railways). The latter tends to increase total mobility, however, which may again increase the environmental problems, instead of reducing them.

Reduction of mobility growth has often been mentioned as another option. Various policies aiming at reducing it can be found in a number of north-west European countries. So far the success has not been very impressive. It appears to be very difficult to curb the mobility drift in an open network economy.

Another policy target is changing the modal split in favour of collective modes. In order to achieve this major investments in rail infrastructure are foreseen, some of which have already been carried out or are in progress. Examples of well-known projects are the Channel Tunnel, the Oresund and Størebelt projects in Denmark, and various high-speed train (HST) tracks all over Europe. But also smaller projects in many countries serve to expand the (capacity of) the existing rail network. This should favour an increase in rail transport; examples of new programmes to increase the volume of rail passengers are Rail 21 in the Netherlands (NS, 1988) and Bahn 2000 in Switzerland (SBB, 1985).

At the European level, it has officially been agreed that, in principle, each transport mode should cover its full social costs, but the actual implementation of this target is being hampered by economic and political concerns. Therefore, transport policies in Europe form a complex ramification of cross-subsidies, hidden subsidies and indirect ways of policy intervention.

3.2.3 The European infrastructure policy

The existence and the quality of the infrastructure is a prerequisite for using vehicles of any mode. In the 20th century, a large road infrastructure network has been constructed in Europe, which has induced a steep growth in the number of vehicles and thereby in personal mobility. At present the road network is relatively dense in comparison to rail or air networks, which partly explains why the car has become the dominant mode of transport.

Table 3.1. CO_2 emission targets in the EU, OECD and EFTA countries

Reduction (%)			Unchanged		Growth (%)
20	Australia	1988-2005	Australia	1988-2000	20 Ireland 1990-2000
20	Austria	1988-2005	Finland	after 1995	25 Spain 1990-2000
5	Belgium	1990-2000	France	1990-2000	
20	Denmark	1988-2005	Greece	1990-2000	
20-30	Germany	1987-2005	Italy	1988-2000	
20	Italy	1988-2005	Iceland	1990-2000	
20	Luxembourg	1990-2005	Japan	1990-2000	
3-5	Netherlands[1]	1989-2000	Canada	1990-2000	
20	New-Zealand	1990-2000	Luxembourg	1990-2000	
			Netherlands	1989-1995	
			Norway	1989-2000	
			Portugal	1990-2000	
			Sweden	1990-2000	
			Switzerland	1990-2000	
			USA[2]	1990-2000	
			UK[3]	1990-2000	

Notes: 1) All greenhouse gases 20-25 per cent
2) All greenhouse gases
3) CO_2, CH_4 and other greenhouse gases

Source: Enquete-Kommission (1994)

Since the 1970s, however, the pace of building road infrastructure has slowed down significantly, especially in the north-western part of Europe, although in the second part of the 1980s many new infrastructure plans were again launched in several countries. The same holds for expanding the capacity of airports (Bruinsma, 1994). In the southern parts of Europe, infrastructure investments are booming, mainly because of economic growth and EU subsidies.

The common EU policy is now aiming to close the gaps in the European (transnational) network ('missing links', 'missing networks' and 'trans-European networks'; see Nijkamp *et al.*, 1994) for both road and rail infrastructure (e.g.

the Betuwe freight railway track in the Netherlands and several HST links). This is mainly done for economic reasons, as building new infrastructure is promoted as a major opportunity for regional economic development (Nijkamp *et al.*, 1994; Rietveld and Nijkamp, 1993). Especially in recent years, infrastructure investments have been used to achieve a competitive advantage *vis-à-vis* other (neighbouring) countries, as more and more other policy fields are covered by common European legislation (Rienstra and Nijkamp, 1997a). Apart from efficiency goals, new infrastructure links are also being built to ensure cohesion in the European space (the equity motive).

It is clear that building more and more (road) infrastructure is increasingly incompatible with sustainability goals, mainly because this growth in supply will soon be followed by additional demand. So, as in many other regions of our world, an ambiguity may be observed in policy-making in the transportation field.

3.3 TRENDS IN EUROPEAN PASSENGER TRANSPORT

3.3.1 Mobility growth in Western Europe

First we present some figures to illustrate recent transportation trends in Europe. From Figure 3.1 it appears that most transport takes place in the European core zone (sometimes known as the 'blue banana') in which most population, transport and economic activities are concentrated (see also Section 3.5.1 and Figure 3.8).

The last three decades have been clearly characterized by a strong growth of surface transport (see Figure 3.2). The mobility growth in land transport was substantial in all Western European countries. It appears to be highest, however, in south European countries like Italy, Portugal and Spain. This may be explained by the fact that the mobility development in these countries before 1965 lagged far behind that of other European countries because of low welfare levels, the lower level of car ownership and the poor quality of the infrastructure.

When the European mobility growth in distinct periods is analysed, we observe that the growth rate in Europe declined somewhat between 1965 and 1985, whereas after 1985 this growth rate returned to its former level (Figure 3.3).

High growth rates after 1985 are found in all European countries, except Austria, Finland and Portugal. Explanations for this rise may be the lower price of petrol after the oil crises of 1973 and 1979 and the economic growth in this period, which increased the number of jobs and the average income of the population (which largely explains car ownership).

It is also clear that the mobility growth in south European countries was especially high at the beginning of this period, while nowadays the growth rates in Europe tend to converge (see also Bovy *et al.*, 1993; Button, 1993).

3.3.2 Mobility growth expectations

Although mobility growth is expected to continue, it seems likely that this will occur at lower growth rates than in past decades. The main reasons are that population growth may stabilize, which may reduce mobility growth. Also the capacity of infrastructure is limited, while unlimited expansion does not seem feasible because of the high external costs involved. Another reason may be that the time budget of individuals for travelling is limited, unless faster modes of transport are introduced. Finally, car ownership levels may stabilize in the next decades, as soon as every individual of 18 years and over has a car (in reality, the saturation level may be lower, as is shown by figures from the Nordic countries).

An example of this reduction may be found in the Netherlands. Between 1965 and 1989, mobility growth was on average about 4 per cent a year. Without strict policy measures the average growth would probably be 2.15 per cent a year in the period 1986-2015, while the new transport policy aims at reducing this rate to about 1 per cent a year (Tweede Kamer, 1990b). However, mobility growth may still be significant, and as a result the transport system may move further away from sustainability criteria.

3.4 THE MODAL SPLIT

3.4.1 Changes in the modal split

Modal split changes have shown a fairly structural development pattern. The growth of mobility was unequal for the distinct transport modes (see Table 3.2). By far the highest growth can be found in air transport; this rise stabilized largely between 1980 and 1985, but again grew very fast after 1985. Also private car use rose very rapidly, while the use of collective modes grew much less, so that the modal split changed in favour of those modes which contribute most to the negative external effects. As a consequence, the modal split of surface transport also changed in favour of the private car (see Figure 3.4). This trend is clear in every country, although in the south European countries (75 to 79 per cent), Austria (72 per cent) and Denmark (79 per cent) the private car has the lowest market share, whereas in the Netherlands, Norway and the UK this share is the highest (86, 87 and 88 per cent, respectively).

The modal share of train traffic is decreasing in every country; in Austria and Switzerland the market share is still the highest (11 and 12 per cent) and in the UK the lowest (5 per cent). Also the market share of bus traffic is declining, although in the south European countries especially this share is still relatively high.

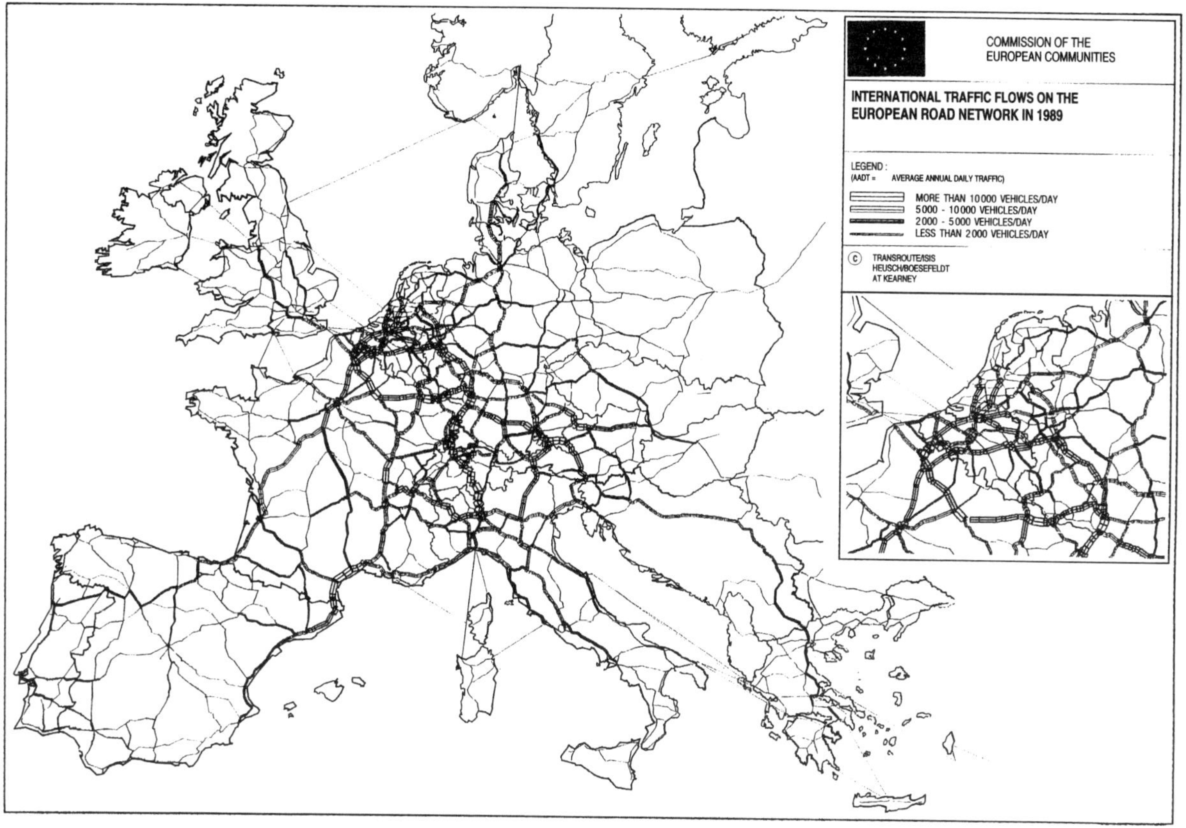

Figure 3.1. Traffic loads 1989 on the European road network
Source: Commission of the EC (1993c)

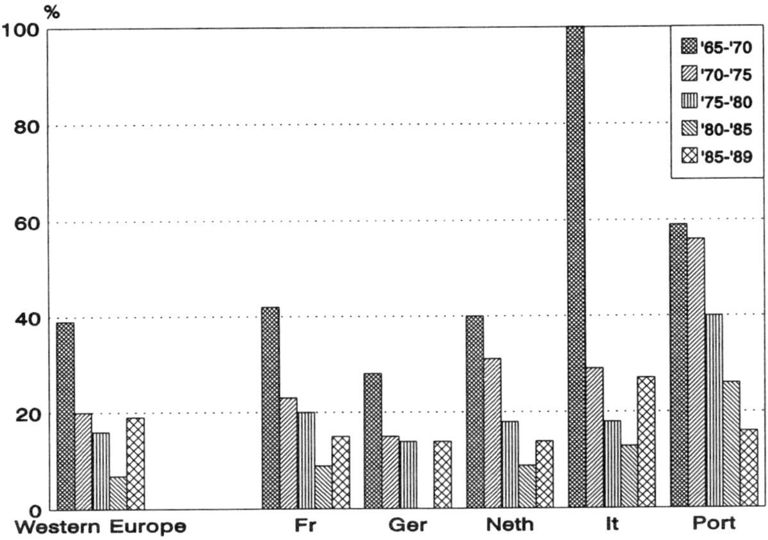

Figure 3.2. Mobility growth in land transport in Western Europe[1,2] (index million passenger kilometres, 1965 = 100)

Notes: 1) Western Europe excluding Luxembourg, Greece and Ireland (because of lack of data)
2) Only passenger kilometres by private car, rail and bus are taken into account

Source: Own calculations based on ECMT (1993)

Table 3.2. Transport growth by mode in Western Europe[1]

	1965	1970	1975	1980	1985	1989
Car	100	151	184	218	237	287
Train	100	104	115	121	126	133
Bus	100	110	126	135	132	138
Air[2]	100	177	288	432	477	702

Notes: 1) The same countries as in Figure 3.2
2) As a proxy for air traffic the number of passenger kilometres flown by the main European companies is used

Sources: Own calculations based on ECMT (1993) (car, train, bus) and Eurostat (several years) (air)

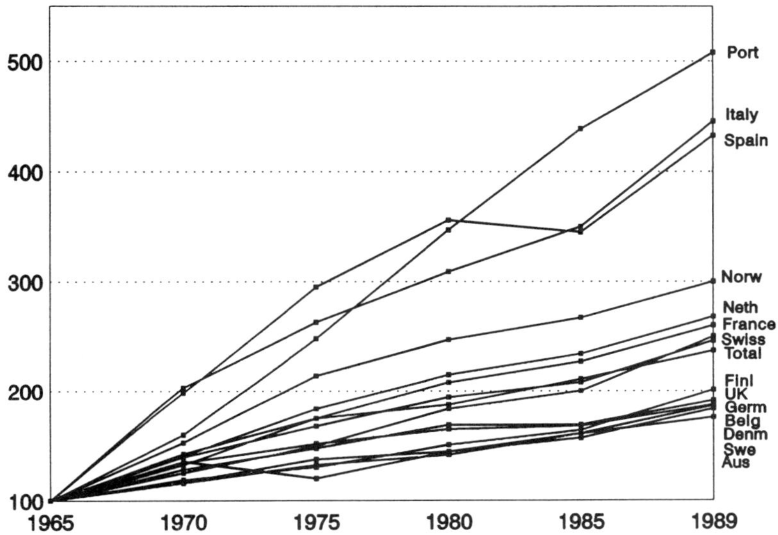

Figure 3.3. Growth in mobility in Western Europe
Notes: See Figure 3.2
Source: Own calculations based on ECMT (1993)

3.4.2 Modal split expectations

The policy of most EU Member States is focused on changing the modal split towards collective modes. It is expected, however, that this policy will probably fail to a large extent if no drastic changes occur. For example, the share of car traffic is so large that even a small change in the modal split will have a large impact on the collective transport system (e.g. a modal split change of 5 per cent decreases car traffic by about 6 per cent, while collective modes will then be faced with a growth in passengers of about 25 per cent). Another example may be found in calculations by the Commission of the EC (1993c) about the effects of a complete HST network on the modal split (see Table 3.3). It appears that its realization decreases the modal share of the private car by only 2.5 per cent, so the car's modal split is still larger than it was in 1987 (see also Hart (1994) for modal split expectations in urban areas).

Interesting reflections on the future modal split can be found in Grübler (1990) and Grübler and Nakićenović (1991a). They argue that every mode has a life cycle. Usually a new mode becomes dominant rather rapidly, but after some time a new competing mode is introduced, and as a result the old mode declines.

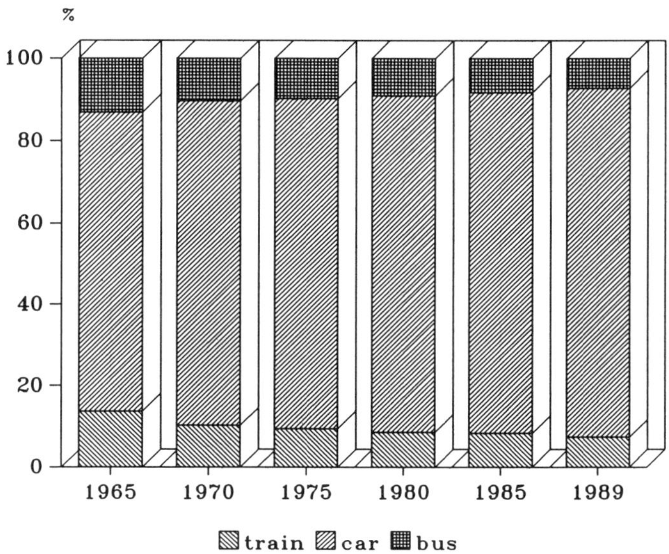

Figure 3.4. Modal split over land in Western Europe
Notes: See Figure 3.2
Source: Own calculations based on ECMT (1993)

The new mode always causes a jump in mobility, so that this rise is exponential. In a certain period even three modes may exist side by side: the growing one, the dominating one and the declining one (see Figure 3.5).

We are currently living in a period in which road transport dominates, while rail traffic is declining and air traffic is growing. According to the life-cycle theory advocated by Grübler and Nakićenović, air traffic will increase its market share, but around the year 2000 a new mode is likely to be introduced, which will start to grow very fast (in the authors' view this may be HST and/or Maglev (electro-)magnetic levitation train).

3.4.3 Transport distances and motives

It is also interesting to analyse over which geographical distances the different modes are mainly used. At a European level there are no systematic data available, however. Hence, we will use Netherlands data in the remainder of this section. In general, these data may be used as a proxy for European developments.

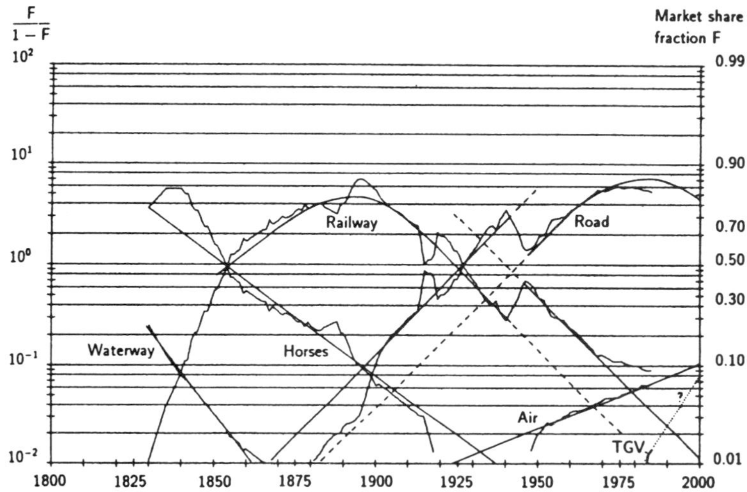

Figure 3.5. Modal split in passenger (intra- and intercity) transport in France

Note: Fractional share of passenger kilometres performed, logit transformation
Source: Grübler and Nakićenović (1991a, p. 23)

Table 3.3. Effects of the construction of an HST network in Europe on modal split

Mode	Base-year	Do minimum[1]		Complete HST network	
	1987	**2000**	**2015**	**2000**	**2015**
Car	78	81	83.5	78	81
Bus	4.5	4	3.5	3.5	3
Rail	13.5	11	8.5	14.5	12
Air	4	4	4.5	4	4

Note: (1) "Do minimum" means that only the HST links which are already under construction at present are in operation in 2015
Source: Commission of the EC (1993b)

It should be kept in mind, however, that the Netherlands has a high density of population (especially in the Randstad) and therefore a quite unique spatial structure.

Another important characteristic is the extensive use of bicycles in urban areas. The travel distances for car and public transport in the Netherlands are presented in Figure 3.6.

It appears that 40 per cent of the trips made by cars are for distances lower than 5 km in the Netherlands. In Europe this share is 50 per cent (Commission of the EC, 1993a). Public transport, on the other hand, is mainly used for longer distances; this holds especially for the train (however, the car is in every distance category the main mode). Also the motives for travelling differ largely for private car and public transport users (see Figure 3.7).

The private car is mainly used for commuting, business and visiting trips; public transport for commuting, visiting and education. The main differences between car and public transport are found in business traffic and education. In the first category public transport is seldom used; for education it is the private car which is hardly used, probably because schoolchildren and students often use public transport (and seldom own a car).

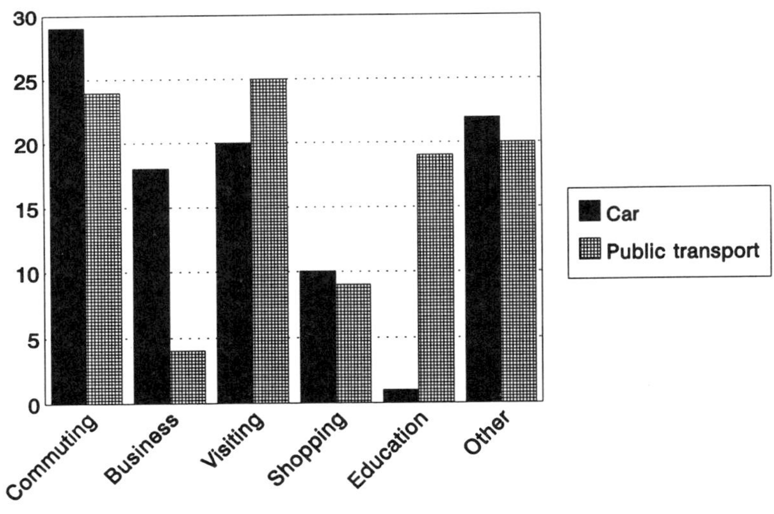

Figure 3.6. The division of trips according to distance category in the Netherlands in 1993

Note: Car: drivers only

Source: Own calculations based on CBS (1994)

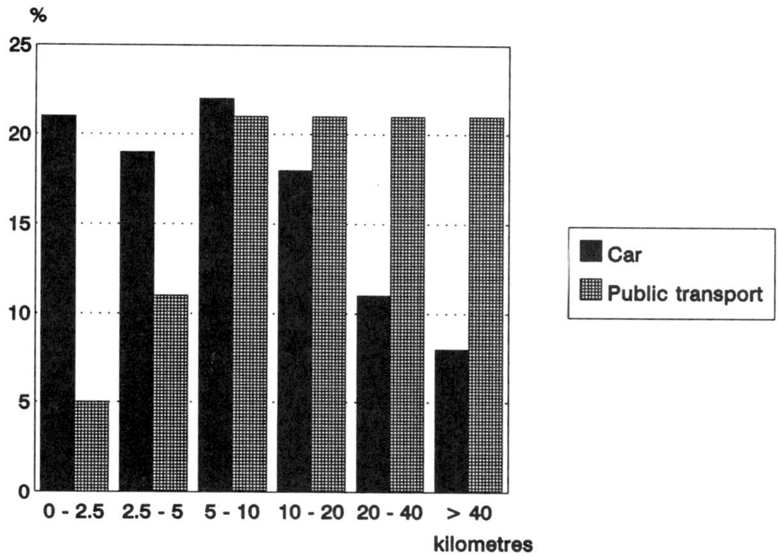

Figure 3.7. Travel motives of car and public transport travellers in the Netherlands in 1992

Note: Car: drivers only

Source: Own calculations based on CBS (1992)

3.4.4 Conclusions

From the available data a high growth rate in mobility can be observed, while the position of the private car has become even more dominant. Furthermore, it should be added that air traffic is growing the fastest. If the emissions of CO_2 and other negative externalities are to be reduced, it is clear that these megatrends are very detrimental for achieving the reduction of pollution.

When the use of the private car is analysed, it appears that it is used mainly for short distances, on which it is difficult for collective modes to compete. Also the motives for using public transport do not give the collective modes much hope of (re)gaining a substantial market share, since it may be expected that commuting and business traffic will rise more than educational traffic (because of demographic trends).

Having now discussed various trends in the transport field, we next move to a discussion of trends in the environment of the transport systems that influence transport behaviour. Four such trends (spatial, institutional, economic and socio-psychological) are discussed in turn.

3.5 SPATIAL TRENDS

In this first section on external megatrends various spatial developments are discussed, namely disparities in regional-economic developments in the EU and future spatial patterns (both supranational and urban).

3.5.1 Trends in regional-economic development at the European level

In the early 1990s, the level of regional-economic development was very different across the EU countries. If the gross regional product of the EU is taken to be 100, regional variation runs from 218 (Hamburg) or 175 (Paris) to less than 50 in parts of Greece, Spain and most of Portugal (Masser *et al.*, 1992). With regard to the future outlook, it is important to consider the origins of these disparities and expected changes. Three factors have influenced the economic success of a region in the past decade:

- **Location within Europe**. Centrally located regions in Europe tend to be more industrialized and economically more successful than predominantly agricultural regions on the European periphery. The European core has taken the form of two different corridors, one from south-east England through the Benelux countries along the Rhine valley to Lombardy, and the second from Paris through north-west Germany to Scandinavia (Masser *et al.*, 1992). These two corridors are served by the highest level of ground transport infrastructure and contain the largest airports. Apart from this reinforcement of European core areas, an additional trend must be mentioned here, that is the increasing importance of 'outliers'. Outlying centres such as Hamburg, Berlin and Vienna, may strongly benefit from the opening of Eastern European markets and the integration of these areas within the Western European economies.
- **Level of urbanization**. Metropolitan areas continue to be more productive and innovative than rural regions, while at the same time the spatial area that is covered by large metropolises has increased (Davelaar and Nijkamp, 1990). In addition, some smaller city regions outside metropolitan areas have also developed innovative economic activities. Thus, national core-periphery divergences still exist, but these have decreased in importance (Geenhuizen and Nijkamp, 1995). Specific border areas may even be in the process of catching up due to their new central position in Europe, for example, based upon their transit function or the establishment of logistic nodes.
- **Industrial heritage**. Regions in which the industrialization was built upon raw materials and heavy industry still suffer from the decline of their major industries. Good examples are north-west England and the Ruhr region in Germany. The high-tech manufacturing and information-based services which

they do attract cannot entirely compensate for the employment loss in traditional industries.

In conclusion, there is a complex number of background factors at work in the European space economy. The question then is whether such factors will change in the future.

3.5.2 Future regional developments

According to Masser *et al.* (1992) there is no reason to believe that the influence of the above factors will change fundamentally in the next decades. One major point of dispute, however, is the spatial pattern within the European core regions. Counter to a trend for concentration of economic activity in metropolitan or non-metropolitan regions ('industrial districts'), there is a trend towards increasingly spread. The former trend is often regarded as closely related to a basic shift in the organization of production, namely from standardized mass production to flexible specialization. The major consequence of this shift is a concomitantly high level of vertically disintegrated and locationally fixed production (Scott, 1988). The latter trend, leading to more diffuse economic development, follows from an increasingly strong footloose character of production, as well as an increased networking of large and small companies on a global scale (Amin, 1993).

Given the previously described basic factors, Masser *et al.* (1992) have envisaged three scenarios in view of the divergent regional economic policy. These are to be considered as extremes and can be summarized as follows:

- **Growth scenario**. Concentration (polarization) will strongly increase, due to factors such as a fierce competition within the EU, additional economic impulses from the expanding Eastern European markets, high-speed transport connections, and the bridging of missing links (e.g. the Alpine crossings). A wide curved zone of intense development will stretch from south-east England across the Channel through the Benelux, the south-west of Germany and Switzerland to Lombardy. In Lombardy, this 'Banana' (Datar/Reclus, 1989) meets a second growth zone, developing along the Mediterranean from the eastern plains of the River Po, then going deep into the Rhône valley in France, and down to Barcelona and Valencia in Spain (the 'Sunbelt') (see Figure 3.8).
- **Equity scenario**. In this scenario, the European regional-economic policy is successful in coping with the above-mentioned trends of concentration. This very interventionist scenario aims at reducing the growing regional disparities. The most remarkable instrument to realize these aims has been created by transforming the European Regional Development Fund into a powerful

agency for investment control. In spatial terms, this scenario leads to a deconcentration of economic growth.

- **Environment scenario**. Here a trade-off has been made between environmental protection and regional economic growth. Some regional industries have been closed down because they impose a too high environmental burden, particularly in Eastern Europe. At the same time, a cleaner and more expensive transport system has influenced the spatial organization of production, i.e. on a smaller spatial scale based upon a larger degree of regional self-sufficiency. On a lower spatial scale level, the aim of urban containment has been given a high priority and has prevented a further spread of economic activities.

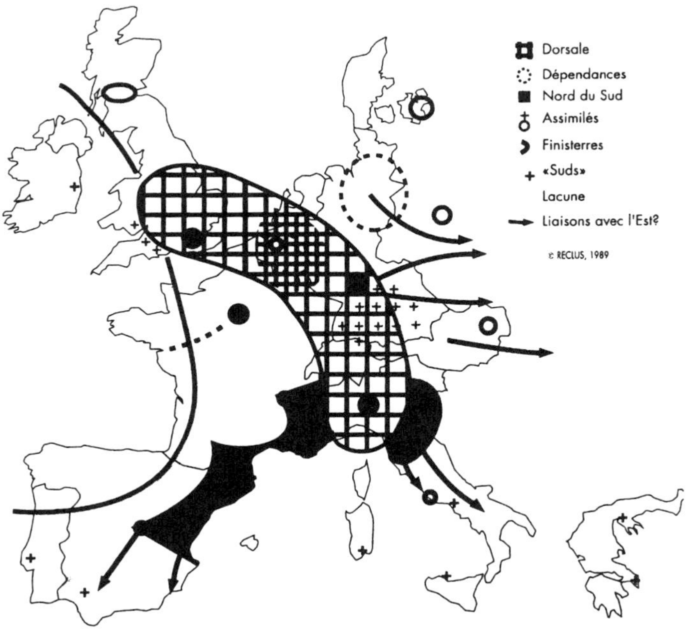

Figure 3.8. The European 'Banana'
Source: Datar/Reclus (1989)

The above scenarios of regional-economic growth clearly have impacts on future spatial structures in Europe. So the next section will focus particularly on these spatial patterns.

3.5.3 Spatial trends on the European level

In general, two differentiating factors in future spatial development can be distinguished, namely (1) the level of concentration (or density) of population and economic activity, and (2) the level of functional homogeneity (or specialization). These two factors clearly form the basis for two extreme models (or contrast images) of future spatial structure in Europe, developed by the Dutch National Planning Agency (RPD (1991)) (see Table 3.3).

The specialization and concentration model articulates a high level of concentration of population, among others as a result of the trend for leading economic (world) functions to settle in top urban locations. Accordingly, the increased internationalization leads towards a hierarchy of functions and a hierarchy of cities, including metropolises at the top, followed by 'europoles' and then smaller cities. This pattern of urbanization is likely to be associated with a hierarchy of transport systems. This will require careful and well-tuned transportation planning.

Table 3.3. Future spatial models at the European level

Features	Specialization and concentration	Chains and zones
Process	Specialization and concentration in large urban centres	Spread over urban regions (possibly some degree of self sufficiency)
Structure	Hierarchy of functions and hierarchy of cities	Flat structure based on footlooseness
Traffic pattern	Hierarchical radial	Criss-cross spatial relations

The other extreme, the chains and zones model, is less oriented towards a hierarchy of functions. Companies in various sectors and at various levels are increasingly footloose, in such a way that the concomitant spatial processes lead towards decentralization. On a regional scale, this model may be associated with some degree of self-containment (self-sufficiency). The chains and zones pattern generally articulates a high-level spread of population and economic activities and is associated with a criss-cross character of traffic and transport relationships. This trend is more in agreement with individual objectives of dispersed residential locations into green areas. If possible, however, a European spatial policy will focus on the connection of urban regions, as well as the bundling of criss-cross relationships in chains.

3.5.4 Spatial trends on the metropolitan (urban) level

The demand for transport is not only a consequence of the relative growth of entire regions, but also of the development of the urban structure within these regions. A general phenomenon experienced by all cities is the suburbanization of living (ECMT/OECD, 1995).

Although the first waves of this suburbanization occurred many decades ago (partly based on public transport), its extent and the scale on which it has developed since the early 1960s have been unprecedented. In particular the private car has brought low-density living within reach of large groups of upper and lower middle-class families. In fact, suburbanization of living was a consequence of various broad changes in society, such as income increase, smaller households, more leisure time and changing housing preferences. Suburbanization is usually associated with negative socio-economic and environmental impacts, such as longer trips to work or for shopping, more energy consumption, pollution, accidents, and problems of public transport provision in low density areas (Masser *et al.*, 1992).

The suburbanization of living was followed by a suburbanization of employment in the 1970s and 1980s. At that time, offices, light industry and retailing started to decentralize, taking advantage of attractive suburban locations with good transport accessibility and lower land prices. Generally, the latter development has caused criss-cross relationships, particularly in commuting. Thus, houses as well as jobs tended to disperse further, in and beyond the metropolitan area, a process which may be called extended suburbanization or counter urbanization (Breheny, 1994).

Decentralized cities are usually negatively valued, because they contribute to an increase in the length of trips to work and prohibit public transport solutions for commuting. However, it should be noted that a few decentralized cities face an interesting development, namely a 'spontaneous' relocation of jobs and houses leading to shorter commuting distances (Brotchie *et al.*, 1994).

At the same time, there are some signs which suggest that suburbanization may have passed its peak and that a phase of reurbanization may have begun. In this view, there is a strong revitalization trend in the inner-city areas, accompanied by the attraction of affluent residents who can afford the increased rents (gentrification). As a consequence, however, low-income inner-city residents are forced to move away from inner-city districts.

Given the above trends and perspectives on urban suburbanization and spread, one particular extreme type of future city can be identified, namely the diffuse (decentralized) city. Large cities, such as London and Paris, and smaller ones, such as Milan and Brussels, seem to conform to this development pattern. But, when spatial planning comes into the picture, a contrasting concept might be taken into account as well. A popular concept in this respect is the 'compact' city, where relatively high-density housing is provided and jobs are contained in the central city and in a limited number of subcentres (Dutch Ministry of Housing, 1991). The compact city has become a leading principle in Dutch physical planning in the past few years, and is currently being adopted in Europe as a guideline in urban planning (Breheny, 1994), although it should be noted that the compact city concept also has some intrinsic limitations in terms of quality of life, land use and prices, and congestion. There is certainly scope for a critical judgement of the compact city idea.

3.6 INSTITUTIONAL TRENDS

Administrative developments proceed in anything but a linear fashion. Long-term stable trends in this context are seldom observed. Through external causes, turning-points appear to occur from time to time, whereby prevailing trends are sometimes abruptly interrupted: from regulation to deregulation, or from centralization via decentralization to recentralization. Indicating long-term trends is therefore an extremely tricky business. In addition, institutional trends may differ from country to country. Nevertheless, an attempt will be made here to call attention to and identify some trends that may have an effect on the development and use of future transport systems. Trends that may be observed at both the national and EU level will be discussed next (see also Nijkamp, 1993).

European integration is a collective term for a series of trends taking place within the framework of European unification. The following impacts may be distinguished:

- **Coordinating policy**. European integration has a great impact on the performance of tasks by national authorities in Europe. Partly in relation to the creation of an internal European market, a large number of policy fields have to be geared together, and if possible harmonized with each other. Transport policy is one of them, but also subsidization policy has to be mentioned. Governments should in principle not grant subsidies that may distort competition.
- **Reducing government deficits**. Within the European Economic and Monetary Union (EMU) certain conditions are laid down for the national economies of the Member States. These 'convergence criteria' relate to inflation, national debt, budgetary deficits and so on. The consequence may be that governments

decide not to perform certain tasks with regard to traffic and transport (e.g. concerning public transport), or to perform them differently.

Other institutional trends concern the use of policy instruments and the role of governments. A number of important trends may be distinguished here (see also Fokkema and Nijkamp, 1994):

- **Decreasing use of the subsidy instrument**. National governments in particular facing budget deficits seem to favour this trend, although it is not clear whether this will be a long-term structural trend.
- **Deregulation in the public sector**. A simplification of rules and regulations facilitates a more decisive development and effective implementation of policy.
- **Privatization of tasks and financing**. In combination with deregulation, privatization in various fields is often seen as an important strategy. A number of tasks, formerly performed by the government, are partly or fully outsourced or left to private enterprises. In transport, the main issues are the privatization of public transport and private financing of infrastructure.
- **Flatter organization with more widespread contract management and self-management**. Although it is not clear whether this trend is to be found throughout Europe, it seems that, in general, a government organization (at the department level) is becoming flatter, with private organizations being utilized for various executive tasks. In addition, more work is visibly being done in contract form ('back to core business').
- **Growth of an agreement culture**. Cooperation between governments and between government and groups in society is being regulated much more by gentlemen's agreements (sometimes known as 'convenants'). Examples may be found in the agreement between government and industry with respect to the introduction of environmental measures and standards.
- **Administrative shifts**. Regional administration is being assigned more municipal and national government tasks (regionalization). Conversely, certain tasks of national governments are shifting to the European administrative stratum (Europeanization). For traffic and transport policy in particular, European legislation and policy are becoming increasingly important.
- **Sustainable development**. In current and future policies of local, regional and national governments, and the EU, more attention will be paid to sustainable development. In various cases 'polluter pays' principles are advocated: through taxation (e.g. ecotax, CO_2 tax) the users of transport systems may have to pay the price for the external costs. These taxes help more environmentally friendly ways of transport to be accepted and adopted. Collective transport systems and alternative energies will then become more competitive.
- **Resistance to infrastructure plans**. In urban areas especially, but also in natural areas, it is nowadays very difficult to construct new infrastructure,

because public decision-making cannot cope in a proper way with social resistance. At present, in various countries NIMBY legislation is being introduced in order to reduce the formal procedures for infrastructure construction. This will not reduce the social resistance, however.

One may thus conclude that institutional patterns of decision-making seem to be going through a phase of structural transition towards decentralization, although it may be added that the transaction costs of this shift are often underestimated.

3.7 ECONOMIC TRENDS

3.7.1 The world economy

Several trends may be observed in the world economy. In the first place, there is a clear trend towards globalization. Therefore, protection of national economies is decreasing (this holds for goods and capital flows as well as for exchange rates), while trade flows are growing much faster than the gross domestic product (GDP). The economies of the Asian-Pacific countries especially have had a very high growth rate in recent decades, and the United States has also enjoyed relatively strong economic development. African and Latin American countries are lagging behind; the same holds to a lesser extent for Western Europe. Finally, one of the major changes in recent years is the shift in Eastern Europe towards free market principles. Such developments may have important implications for transport and the achievement of sustainability criteria (Hart, 1994).

It is difficult to predict future developments of the economy, since such a prediction is dependent on many unknown factors and variables. Therefore, one often resorts to scenarios of a macroeconomic or even global-economic nature. Four such scenarios have been constructed by the Dutch Central Planning Bureau (CPB) for the world economy with different trajectories, which run until 2015 (CPB, 1992a). The indicators of GDP growth and the expected CO_2 emissions are given in Table 3.4.

The first scenario is called *global shift*. In this scenario, the United States shows a positive development because of the competitiveness of the economy, and likewise the Asian-Pacific region, mainly because of the liberalization of the markets in this region. The relative economic position of Europe declines, however. This happens because no serious measures are taken to liberalize markets, the EU is split up and every country takes its own (often protectionist) measures. Also, because there is insufficient cooperation between these countries, no sufficient and efficient environmental measures will be introduced.

The second scenario is called *european renaissance*. In this scenario important economic trade blocks develop, which are supported and protected by the governments concerned. The EU develops towards integration, e.g. by completing the EMU. In this scenario the European economy develops relatively positively, while the USA lags behind. In Europe a CO_2 tax is also introduced and the aim of the European Energy Charter is achieved. This introduces low-emission Russian gas to the European countries, which has a positive influence on the CO_2 emissions in Western Europe.

The next - somewhat pessimistic - scenario is called *global crisis*, in which Europe and North America lag behind, while Japan and other Asian countries have a relatively high growth. A negative spiral develops, which finally leads to a shock, followed by a deep economic recession. No serious environmental policy is introduced; as a result the CO_2 emissions grow rapidly, despite the lower growth rates.

The most positive scenario is *balanced growth*, in which the world tries to reach a situation of sustainable economic development. There is much cooperation in the world, protectionism is abolished and a worldwide CO_2 tax is imposed, for which the developing countries are compensated. Therefore, GDP rises, while the CO_2 emissions are largely decreasing all over the world.

Table 3.4. GDP growth and CO_2 emissions in the CPB scenarios (1990-2015)

	WLD[1]		NAM		WEU		JAP	
	DP[2]	CO_2[3]	DP	CO_2	DP	CO_2	DP	CO_2
Global shift	3.4	163	3.4	144	1.9	108	4.3	156
Eur. renaissance	2.9	143	1.8	116	2.8	103	3.7	138
Global crisis	2.2	151	1.7	122	1.8	117	3.0	139
Balanced growth	3.6	74	3.0	77	3.2	84	3.1	93

Notes: 1) WLD = world, NAM = North-America, WEU = Western Europe, JAP = Japan
2) Gross Domestic Product, average annual growth
3) Index of emissions: 1990 = 100

Source: CPB (1992a)

In most scenarios it appears that CO_2 emissions are rising; only in the balanced growth scenario is this not the case, despite the fact that in this scenario the highest growth in GDP is achieved. From these scenarios it may be concluded that CO_2 emissions can only be reduced by worldwide coordination to solve this problem. In another study, the CPB (1992b) has also calculated the consequences for Dutch transport with respect to each of these scenarios (except global crisis).

3.7.2 The European economy

In Western Europe there has been a clear trend towards economic integration since 1945, mainly by means of the EU (and its predecessors) and the European Free Trade Association (EFTA). Back in the 1950s the EU started with cooperation on some strategic fields (coal, steel) among six countries; slowly it was extended to other markets and more countries (to 15 in 1996). In 1992 a Single European market was finally introduced, with (in principle) free flows of goods, labour and capital. Also many regulations and laws have been standardized. It is foreseen that by the end of the century the EMU will also come into being. At the moment there is still much opposition to this plan, however.

The Dutch CPB (1992a) characterizes the European economy as follows:

- most stress is placed on coordination principles instead of free-market principles;
- there is an extensive paternalism in numerous fields of government policy;
- in economic policy there is a strong emphasis on income distribution;
- there is a strong drive to reduce risk and uncertainty, which causes many rigidities, weakens incentives and hampers individual adaptability and inventiveness.

Because of these characteristics the development of the European economy has lagged behind that of the United States and the Asian-Pacific countries in the last few decades, which has also led to high unemployment rates in Europe. There is, however, a growing attention to free market principles here and worldwide. It is questionable, however, whether this is sufficient to compete with other parts of the world (see the scenarios in Section 3.7.1).

3.7.3 Economics of the transport sector

In the nineteenth century the transport sector was largely privately financed and operated, but later on governments took responsibility for the financing and operation of infrastructure. For example, the railway companies were nationalized and later faced huge losses (Rienstra and Nijkamp, 1997b).

In recent years profound changes in economic and spatial policy have brought about a re-orientation in transportation with a clear focus on a market organization, in which the role of the public sector is increasingly no longer taken for granted. The trend towards market principles and liberalist views has caused trends towards deregulation, decentralization and privatization (Button and Pitfield, 1991). More emphasis is put on the profitability (or at least cost coverage) of transport. This holds especially for public transport, where subsidies have been cut, tariffs have been raised and the companies are made more independent from their governments. For example, bus and railway companies have obtained autonomy or are even privatized in Sweden, the United Kingdom, the Netherlands and other countries. Also schemes to measure the social and economic costs and benefits of infrastructure projects have become increasingly popular in many countries.

Another issue is the liberalization of the transport market in the EU. Regulations have been standardized and the markets for goods (for example, cabotage) and rail transport have been, or will be, liberalized (Wit and Van Gent, 1996).

For budget and environmental reasons, most European countries have raised taxes on gasoline and other conventional fuels in such a way that they dominate fuel prices. Furthermore, in every country discussions are taking place about levying road tolls or introducing systems of road or cordon pricing.

In conclusion, both from a macroeconomic and a transport-economic perspective many changes are likely to occur, which will have profound impacts on the transport sector.

3.8 SOCIO-PSYCHOLOGICAL TRENDS

3.8.1 Introduction

Society is permanently subject to changes, as a result of many trends both small and large, temporary and long lasting, which take place in all sectors of society. This may appear trivial, but a large number of these alterations may directly or indirectly influence people's travel behaviour. Therefore, insight especially into socially broad trends is relevant in order to understand mobility developments in the past as well as to anticipate future tendencies (see also Marchetti, 1994; Salomon *et al.*, 1993).

In general, social and cultural trends will influence travel behaviour and travel mode choice to the extent that these trends are linked to the determinants of individual behaviour. This means primarily that social changes which affect (perceived) travel mode options and travel needs, can - and probably will - lead to behavioural adjustments (Van der Hoorn, 1989). Whether at the same time the travel mode choice changes, will depend on the effects on existing preferences, norms, attitudes and habits (the so-called ('personal motivators') (Carlson *et al.*, 1982). In this way discrepancies may occur between public and individual interests. Only when environmental problems become very serious in the opinion of individuals, do they become serious social issues that may become dominant in the decision-making process (e.g. the rise of the green movement).

Important factors determining individual travel mode options and travel needs include among other things, the composition of the household, the place of residence, occupation, lifestyle, income and activity patterns (Van der Hoorn, 1989). So when alterations take place with respect to one or more of these factors, eventually the travel behaviour may change. As it appeared above, this also holds true for alterations as a result of economic and institutional trends and trends in the fields of spatial organization and technology. It is clear that certain social changes can only take place in the context of developments in other fields.

In this section a few relevant trends are briefly described. In addition, probable consequences for travel behaviour and travel mode choice are presented in qualitative terms.

3.8.2 Increase of leisure time

During the present century working hours have been considerably reduced. Whereas in the last century the common working week amounted to approximately 50 hours, nowadays the working week is 36-40 hours. Moreover, this reduction seems set to continue in the next few decades; the possibility of restricting the working week to 32 hours is under discussion within the framework of improving the distribution of available work. At the same time, housekeeping tasks have become far less time-consuming as a consequence of the availability of electric household appliances, smaller households and the increasing use of ready-to-use products. Because of these developments the amount of leisure time has greatly increased and will probably continue to do so in the future (Leidelmeyer *et al.*, 1993).

The increase of leisure time has a large influence upon people's activity patterns and upon the way they spend their time indoors, as well as outdoors. As a result of this, travel needs will change; the need for work-related travel may slightly decrease, while the need for social and recreational travel will probably increase substantially. An illustrative example in this respect is the recent trend towards several short holiday trips per year in addition to the 'traditional' summer holidays.

Regarding travel mode choice, the preference for the individual transport mode (private car) in particular is expected to rise while the use of collective transport modes will decline, at least without clear intervention measures. This may happen because for social and recreational travel more importance is attached to flexibility and the possibility of taking along luggage. Moreover, public transport options are generally more restricted for such trips and frequently private car trips are made by more than one person, which adversely affects the competitive position of the collective transport mode options.

3.8.3 Ageing

As a result of lower birth rates, better health care, and more healthy living conditions and diets, the proportion of elderly people in the population is increasing. In addition, older people tend to be independent for a longer period and, as the average age of retirement has been lowered, there is an ever-growing group of still energetic people with a large amount of leisure time at their disposal. These developments probably also affect their lifestyle as well as their activity patterns to a large extent and as such influence their travel needs and travel behaviour, car ownership rates and so on (Büttner and Grübler, 1994). Work-related travel may cease for them, but more time will be spent on social and recreational travel.

Hand in hand with this trend the preference for individual transport is also expected to increase, partly as a result of the increase of driving competency among elderly people. Moreover, this group has been accustomed to this mode of transport more than the elderly people of the recent past.

3.8.4 Individualization

Probably one of the most significant trends of the last century is individualization (Jansweijer, 1987). This trend, which occurs in almost all segments of society, implies that the individual is increasingly placed in the centre. Ties are becoming weaker with both the direct social environment and the groups to which a person belongs. This is true for the individual who is less led by group standards in his or her way of thinking and acting, as well as for the community, which treats the person more as an individual and imposes less stringent group standards. The reverse of the medal is a reduction of the society's opportunities to influence individual behaviour, however.

In connection with these general individualization trends, there are a number of more specific interrelated trends, which lead to a major change of living patterns. Firstly, 'household slimming' is a major trend. Over the last few decades there has been a sharp increase in the number of one- and two-person households (Batenburg and Knulst, 1993). Moreover, as a result of birth control, family size has strongly decreased during the last 50 years. Secondly, there is a growing participation of women in the workforce. More and more women work for a longer period of time. As a result of this, the age at which women have their first child is increasing (on average worldwide, this age is the highest for Dutch women). But also after having children many women keep their jobs. Both trends coincide with a third one: the increase in task combination. Due to the smaller households and the growing labour participation, an increasing number of people have the opportunity and wish to combine several tasks (household, family care and work). As a result of this the traditional male and female roles are becoming more and more blurred.

It is clear that all these developments imply radical changes for the activity patterns and travel behaviour. It is expected that these developments will continue in the next decades. The result is a strong increase in travel mode needs with possibly, especially for task combiners, a high preference for fast and flexible individual transportation. We may conclude that individualization is likely to reinforce the so-called 'prisoner's dilemma': individuals recognize the importance of a clean environment, but are unwilling to accept the consequences in terms of individual mobility (see also Section 4.6).

3.8.5 Globalization

Another important trend with far-reaching implications and consequences for transportation is globalization. In many fields, such as trade and economy (see Section 3.7.1), politics, science, technology and tourism, a strong sense of internationalization has developed during the last century. Developments in telecommunication have encouraged most countries to become much closer. Communicating with someone on the other side of town is, apart from possible language barriers, hardly different from communicating with someone on the other side of the globe (the 'digital highway' may be the next trend in this context). As a consequence, in a way the opposite tendency to individualization is gradually coming into existence: at the global level an increasing social cohesion seems to be emerging (the 'global village').

Together with the introduction of fast and safe travel options for long distance transport (plane and high-speed train) a growing need for international and intercontinental trips is arising.

A trend which may partly be related to the 'globalization' trend is the growing concern for the environment. New developments in the fields of science and technology have generated increased insights into climatological and environmental processes and the effects of human behaviour on the ecosystem. Partly as a consequence of this, there is a growing concern for and real worry about environmental issues, which, among other things, leads to the need for international cooperation for research into causes and possible solutions of the problems.

In view of the fact that the area of traffic and transport is one of the main culprits in the environmental issue, a growing sense of environmental awareness can play an important part in the social and political acceptance of mobility control measures (Rooijers, 1992b). So far, this has not yet significantly led to a relevant change in travel behaviour and travel mode use, mainly because the competitive position of alternatives that are less harmful to the environment is limited for current transport systems. In addition, many people feel that the environmental issue is mainly a technological problem (Levy-Leboyer and Duron, 1991).

3.8.6 Levelling

Finally, a trend which may be less manifest than the previous ones, but which is in fact equally important concerns levelling of social classes because of the growing attention to equity in Western societies. The sharp dividing lines of the past between the upper, middle and lower class have become less and it is the middle class especially which is becoming the ruling social class. As a result of this development, which has occurred mainly in recent periods of economic growth, car ownership and use have come within virtually everyone's range. Thus the car has become deeply anchored in the lifestyle and activity patterns of a large group of people. In particular, price measures, which would make the ownership and use of the car financially prohibitive for a large group, are considered inequitable by many people.

In general, it may be concluded that most trends in the socio-psychological field reinforce a lifestyle characterized by mobility growth and a modal shift towards the private car and plane.

The trends sketched above seem to be plausible in many European countries and are based on various background studies. Nevertheless, it is a fascinating question to consider whether transportation experts share the view that these commonly accepted trends will come to fruition during the first half of the 21st century. This is dealt with in the next section.

3.9 EXPERT VIEWS ON EUROPEAN TRANSPORT TRENDS

3.9.1 Introduction

After the previous mapping of current trends in the transport sector in distinct related fields, it may be interesting to analyse how experts value these trends, *inter alia* from the viewpoint of realism and sustainability criteria. In this framework an extensive investigation has been made of the views of transport experts in the Netherlands, followed by a similar review of expert opinions of various European transport researchers. The Dutch transportation experts were taken from the membership of the Dutch *Colloquium Vervoersplanologisch Speurwerk* (CVS), a scientific organization comprising most Dutch transport researchers. The European transport experts were selected from the Network on European Communications and Transport Activity Research (NECTAR), a scientific body encompassing many European transport researchers from most European countries. Here we mainly present the results from the investigation among Dutch experts, based on a postal questionnaire sent to hundreds of transport specialists (sample size 752; response rate 36 per cent). The survey was carefully designed and tested before it was circulated. It contained a wide spectrum of questions on perception, expectations and desirabilities of sustainable transport policies at present and in the future. Expert answers on a long list of questions could be given by using qualitative or quantitative scores. The main aim of this experiment was to construct future scenarios, which will be discussed in later chapters.

The questionnaire was split into three sections:

- **background** of the respondent;
- **general views** on trends and on expected and desired situations in the year 2030 in various fields related to transport;
- **specific views** about expected measures and mode changes in (sustainable) **transport systems in 2030**.

Here we will present the results of the questionnaire, as far as they relate to views on megatrends. Other findings will be presented in Chapter 9. First, a short introduction is given to the questions on trends including the reason for using them in the questionnaire. Then the answers and scores given are presented. Finally, some interpretative conclusions are drawn. The results are presented in three separate subsections.

3.9.2 General views of the experts on trends in transportation

The general opinions about trends in the transport sector mainly concern questions about whether the major trends in transport will continue and about expectations as to whether the transport system may be sustainable in 2030.

More than 80 per cent of the respondents are of the opinion that the current trends in transport cannot be maintained (see Figure 3.9). Only 15 per cent think that the transport system will be sustainable in 2030, while more than 80 per cent are more pessimistic about this, answering 'no' or 'may be'.

Therefore, it may be concluded that, according to most experts, changes in the transport system are necessary and unavoidable, a finding which is perhaps more common wisdom than striking. In general, however, it appears that Dutch transport experts are cynical or even pessimistic about the possibilities of altering these trends sufficiently. Apparently the effects of transport policy and other (e.g. technical) developments are not regarded as sufficient to achieve a sustainable transport system, which is in fact a rather discouraging conclusion.

3.9.3 The perceived relative importance of transport externalities

The experts were also asked to give scores on a 1-10 scale to several environmental externalities caused by transport. In this way it is possible to analyse which externalities are thought to be most important and which changes are necessary for achieving a sustainable transport system.

Secondly, the relative importance of various background factors - like climate change - may then be compared to other externalities caused by transport.

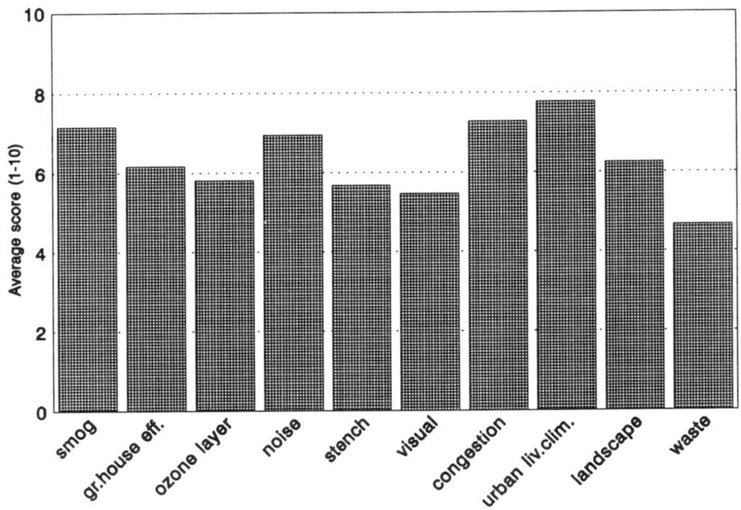

Figure 3.9. Expert opinions about trends in transport
Note: $n = 271$

The average scores given to the various externalities distinguished are pre appears clearly that according to the experts, transport causes most problems in the urban (living) environment, followed closely by congestion, local/regional air pollution and noise annoyance, while by far the lowest score is given to waste problems. Other low scores are found for intersection of the landscape (spatial segmentation), stench annoyance and visual intrusion. The global environmental problems of the greenhouse effect and the depletion of the ozone layer are given relatively low scores too. Other externalities mentioned in addition to the ones listed in Figure 3.10 are road accidents (29 times, with an average score of 8.00), waste of non-renewable resources (7 times) and the need for scarce space (3 times).

It may be concluded that especially the externalities that directly affect people - which are found at a local and regional level - are thought to be the most important problems. Among these problems stench annoyance and visual intrusion are thought to be less important, which may be the result of the current Dutch policies aiming at reducing these types of nuisance by discouraging the use of cars in city centres and the construction of noise shields around highways and railways in urban areas.

What is striking is that congestion is found to be a major externality, although the results may be influenced by recently renewed discussions in the Netherlands about this subject. It appears that congestion is seen as a bigger problem than many environmental issues, although congestion may also contribute to environmental problems because it leads to a larger energy use. Intersection of the landscape is given only a moderate score, although in the past few years there has been an intensive discussion in the Netherlands about the new Betuwe freight rail track to Germany and the HST connection between Brussels and Amsterdam, two infrastructure provisions which will slice through rural areas and nature parks.

Global problems are apparently given relatively low scores, which may reflect society's limited awareness of the wider world. It is interesting, however, that the standard deviations of the greenhouse effect and ozone layer scores are the highest, which may be an indication of the fact that the experts disagree most on the scores for these problems. Nevertheless, in general it appears that such problems are not seen as the main threats to the environment.

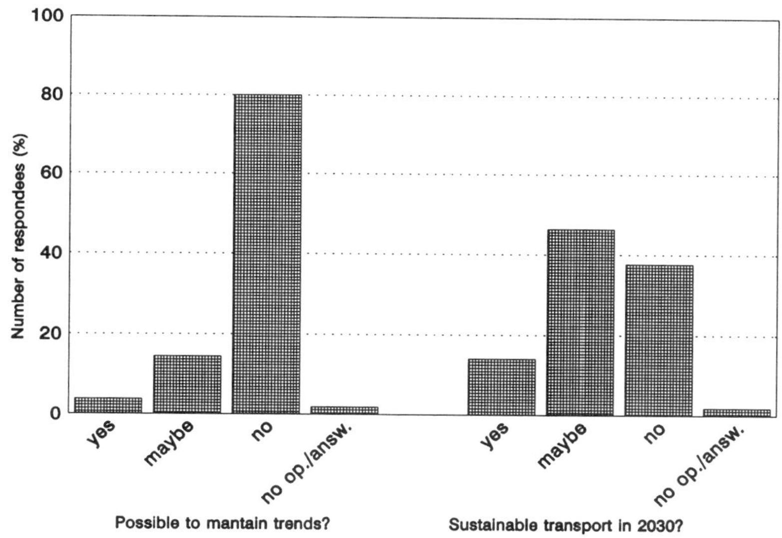

Figure 3.10. Average expert scores (1-10) of externalities caused by transport

Notes: 1) Scale 1-10; 1 is lowest, 10 = highest
2) n = 266-269

3.9.4 Expected consequences of various trends on transportation

Expert opinions were also collected on the expected effects of trends in the institutional, economic, socio-psychological and technical fields. A brief summary of answers on the issues related to the economic and institutional fields is presented in Figure 3.11. At this stage, no questions were asked about *spatial trends* because the future spatial organization is used later on as the basis of the scenarios.

The deregulation of the transport sector appears to be one of the main trends in the *institutional field*. This impact appears to be perceived negatively for sustainability, since 56 per cent of the Dutch experts answer that deregulation is a moderate or substantial obstruction for achieving this aim.

In the *economic field* the attention was *inter alia* focused on the question of an increase in taxes levied on conventional fuels (for example, a doubling of the price). This will certainly have an impact on car use, but this impact will only be modest according to the majority. Almost 50 per cent, however, expect the impact to be big or even very big. The impact of an increase in the price of fuels on the use of collective modes will be much lower, however. This impact will only be moderately positive, while several respondents give the answer 'small' more often than 'big'. So an increase in the price of fuels tends to reduce car use quite substantially, but the shift in the modal split towards collective modes is expected to be relatively small. In most cases collective modes do not appear to be a viable alternative for the car, so that the impact of a drastic pure fuel price increase may imply a reduction in the overall mobility level.

Mobility growth is also thought to be an inevitable consequence of economic growth in the future. A linear relationship between these two growth variables is very likely, a trend which is about the same as in the past (Nijkamp, 1994). It is interesting to note, however, that a small majority of 52 per cent think that it is possible to maintain economic growth with a lower or even negative mobility growth rate (29 per cent think that no further mobility growth is necessary for maintaining economic growth, while 12 per cent thinks the level may even be reduced). It is plausible that this group favours a policy of 'selective growth', which means that environmentally harmful sectors of the economy are not allowed to grow (or are even forced to halt their activities), while environmentally more friendly sectors are stimulated.

The experts' views on the socio-psychological and technical aspects are presented in Figure 3.12. One of the main trends in the *socio-psychological field* is the trend towards individualization. This will have a big negative influence on the possibility of achieving a sustainable transport system, probably because of the increase in smaller households (and, as a consequence, construction of more dwellings and, hence, a more diffuse spatial structure) and the emphasis on individual - more polluting - travel modes.

About 51 per cent of the respondents perceive as moderate the willingness of individuals to sacrifice part of their mobility in order to achieve more sustainability; about one third thinks this willingness is small, however.

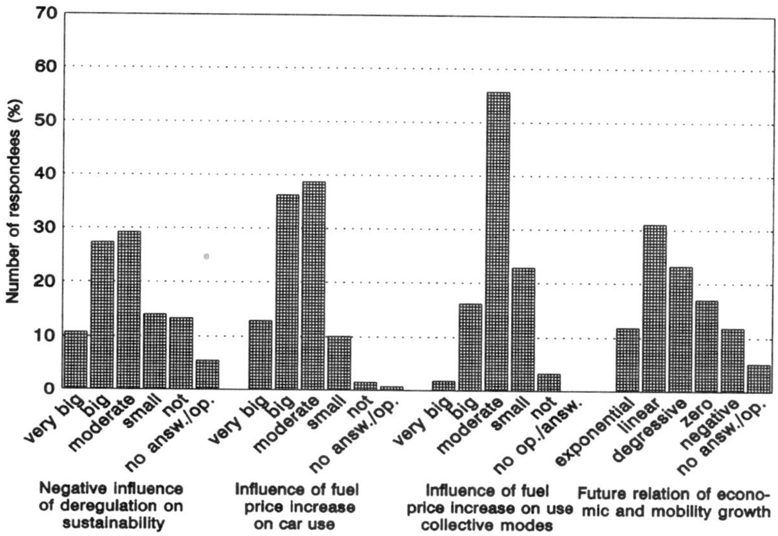

Figure 3.11. Expert answers given in the institutional and economic field
Note: $n = 271$

Therefore, there is not much hope for the acceptance by individuals of the introduction of counter-mobility measures in the transport system, although it may be generally assumed that these sacrifices refer to externalities directly affecting people (mainly at a local and regional level) and not to global problems like the greenhouse effect.

Technical developments like teleconferencing and virtual reality are expected to have only a moderately negative impact on the mobility level, while a large minority sees these technological advances as having only a small impact on achieving a more sustainable transport system.

In conclusion, it appears that in the institutional, economic and socio-psychological field most trends do not favour the achievement of a sustainable transport system. Deregulation and individualization will have a negative influence on sustainability, while economic growth will give a major boost to mobility growth. Increases in fuel prices may have a moderate impact on the choice of transport modes. The results found here are largely consistent with the findings presented earlier in this chapter. The confidence of experts in the effectiveness of policies is clearly not overwhelming. They seem to presuppose a degree of inertia in public attitudes, a finding which may also be found, for example, in the Annual UK LEX Monitoring Reports.

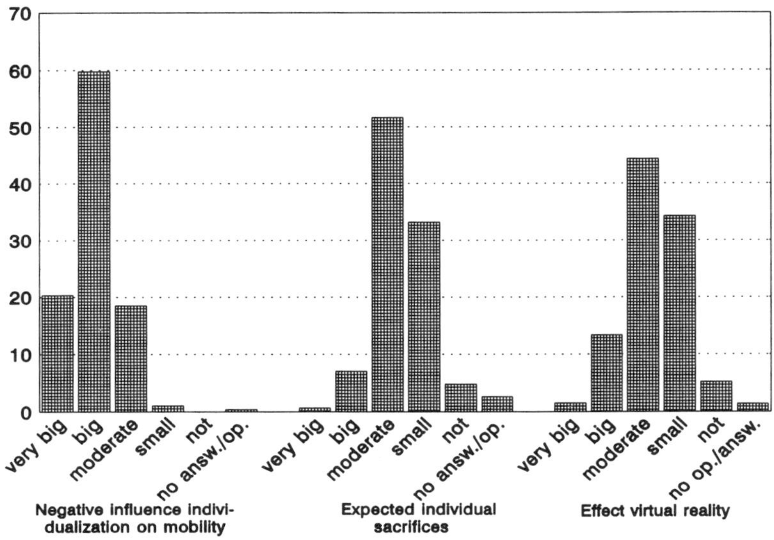

Figure 3.12. Expert answers given in the socio-psychological and technical field

Note: $n = 271$

The results from expert views do, however, offer some cautious hope, in that large minorities are still relatively positive regarding the impacts of these trends and the possibilities of achieving a (more) sustainable transport system.

3.10 CONCLUSIONS

In general, it may be concluded that the current major trends in transport tend to move the transport system away from sustainability, despite the efforts in the technical field to reduce the negative environmental impacts of transport. Mobility growth is expected to continue, together with an increasing share of the private car and the plane in the modal split. This trend is reinforced by trends in other fields, which appear to have a negative impact on the sustainability of the transport system in general.

From this first observation, it may be concluded that a more drastic change is necessary, as otherwise the prespecified environmental goals will not be achieved. Such a change might be initiated and stimulated by governments, since forces in society will not autonomously lead to a sustainable transport system.

There are several ways to reduce the external impacts of transport. Mobility levels can be reduced, a modal shift towards more environmentally-friendly modes (mostly collective ones) can be stimulated, current transport modes can be improved, new - more environmentally friendly - transport modes can be introduced, etc. These possibilities can be achieved by many types of policy, e.g. spatial policies, influencing the behaviour of individuals, pricing measures, construction and provision of new transport modes, etc. In part B, we will discuss the potential of these possibilities by means of an analysis of (1) *background factors* influencing the mobility level and the modal split, (2) the *success and failure factors* of new and collective transport modes and (3) the *potential of new technologies* and transport modes.

Part B

ANALYSIS OF NEW TRANSPORT SYSTEMS

4 NEW TRANSPORT SYSTEMS: SUCCESS AND FAILURE FACTORS

4.1 THE POTENTIAL OF TRANSPORT TECHNOLOGY FOR SUSTAINABILITY

The analysis in the first chapters has demonstrated that the transport system is clearly moving away from the achievement of sustainability criteria. It may therefore be attractive to solve the environmental problems of the transport sector by resorting to technical solutions, since these tend to generate the least resistance in society. Two major directions in technology may be distinguished: improvements of current modes and the introduction of new ones.

Technology itself is fairly neutral; it is its use which determines its environmental merit. Technology may contribute to the attainment of sustainability standards in four ways:

- it may give society the means to 'clean up the mess' already produced (abatement and removal technology);
- it may provide environmentally improved means to perform a certain activity using prevention technology (e.g. the catalytic converter);
- it may increase the attractiveness of less polluting activities in transport this may lead to a substitution between transport modes;
- it may supply substitutes for activities which cannot be undertaken in a cleaner way with the given technology an example is teleworking.

In assessing the impact of new transport technologies, one should recognize the fact that transport systems (broadly defined as the interaction between vehicles, infrastructure and its management) are never static, but always in a state of flux. In a dynamic environment developments in vehicle technology, infrastructure technology and systems technology may gradually change the environmental profile of a specific transport mode, in both positive and negative directions. Thus, the environmental profile of a given transport mode (system) is a variable indicator, which may change over time. Initially, environmentally-friendly transport modes may lose part of their advantages if compared to improved, formerly less benign, modes of transport. Clearly, entirely environmentally friendly transport systems do not exist; there are very clean modes of transport (bicycles and walking), however.

A qualified and comprehensive assessment of the environmental impact of transport systems should theoretically be based on the full life cycle of vehicles. Since many of the data needed for a life-cycle analysis are either unavailable or inaccessible (see e.g. Adriaanse, 1993), a comparative analysis of the environmental impact of transport modes has many 'blind spots'. In most cases rather general data for specific environmental impacts are available, whereas site-specific data are needed. It may be assumed, for example, that negative externalities are higher in densely populated areas than in rural areas. Neither are there sufficient data available on the cumulative effects of environmental pollution. This also hinders a balanced environmental data analysis in the transport sector. It is therefore often necessary to use segmented data or simply stick to a qualitative analysis.

Nevertheless, the theory of the life cycle of technologies presents the various stages in which a transport mode may function. Therefore, this topic will be elaborated first.

4.2 THE LIFE CYCLE OF TECHNOLOGIES

The life of a transport system - like almost any other product - may be described graphically by an S-shaped curve. This curve represents the social acceptance (or diffusion) of the product (see also Figure 3.5). This diffusion process consists of four stages; emergence (birth), growth (market acceptance), maturity (market saturation) and decline. These four stages will next be elaborated (based on Geerlings, 1995; Grübler, 1990; Marchetti, 1987; Nakićenović, 1991; Stam, 1991).

4.2.1 Emergence

During the lifetime of a certain transport mode new modes may be invented which compete with existing modes. 'Often, new technologies create "niches" that lead to products and services hitherto unavailable. More frequently though, successful innovations can fill an established niche by providing improved technical and economic performance or the social acceptability of existing services through new ways of fulfilling them' (Grübler and Nakićenović, 1991b). Niches in this respect may be defined as 'new ideas set in motion or introduced to solve urgent problems', whereas the term 'invention' may be defined as 'the discovery of a potential solution' (Luger, 1991).

Innovations may be subdivided into product and process innovations. Product innovations are typically aimed at opening new markets, both in positive ('pull') and negative ('push') marketing strategies. They are most likely, but not uniquely, found in the earlier stages of the product life cycle. Process innovations are frequently part of cost-reducing strategies, and will be found most frequently in later stages of the product life cycle. Both types of innovations may be used to extend the life of a product (or its producers).

This evolutionary substitution process changes the structure of transport systems. These new modes also show this typical lifetime pattern. Between the starting-points of the long waves of transport modes a period of at least 55 years is commonly mentioned in practice.

This technology wave does not only resemble the Kondratieff wave, well known from economic growth theory, but may also be explained by it. Economic theoreticians like Schumpeter (1939) more or less 'predict' that at the end of a certain period only a smart combination of new technologies (*Neue Kombinationen*) are able to escape the 'Kondratieff barrier (tunnel)' and then give a strong boost to a slackening economic growth regime (crisis). They can only do so, however, if their predecessors disappear, often violently. The fact that a certain technology has a long history, strongly reduces its potential to keep up with newly invented technologies. In other words, existing systems reflect 'the state of precursor technologies at the time of their birth, the nature of markets at that time, and factor prices at the time'; so 'if the system was invented today, it would be different' (Garrison, 1989, p. 108).

This (violent) substitution process is also called 'creative destruction'. The life cycle of transport systems is relatively long, sometimes even 100 years or more, covering two or more Kondratieffs. Finally, it should be added that each transport system introduces its own environmental disadvantages, while eliminating those of its predecessors (Grübler and Nakićenović, 1991b).

It should be recognized that it may take a very long time to develop and introduce new technologies; this situation is called 'technical inertia'. An example may be the Maglev high-speed train (although also other factors may have played a role here). A test track was constructed already in 1962, but it is expected that the first train will not run before the year 2000 (a network will - if ever - be constructed much later). Therefore, entirely new modes may not provide many, if indeed any, opportunities to solve the environmental problems caused by transport, if introduction periods are not accelerated.

4.2.2 Growth and maturity

Growth and maturity are the next two stages in the market penetration of a new technological system. In the transport sector, this has far-reaching implications for the modal choice and spatial action radius of travellers. Users of new transport systems are not only able to travel faster, but they may also use a far larger (10 times or more) infrastructure network than they were used to. This leads to a hierarchy of territories for each mode ('boxes-within-boxes architecture') for a given time-distance horizon. People allocate a given part of their time budget to their daily travel needs, which is fairly constant over time

(Salomon *et al.*, 1993). Yet, faster modes may enable them to cover a greater distance over the same time.

These diffusion and substitution processes are not evenly spread between countries (regions); there are early adopters and followers, which points to an international bandwagon of technological innovations. Countries which are early adopters usually have a larger network with a higher level of service, but the diffusion process is much slower with a longer learning and construction phase than that found in countries which are reckoned to be innovation followers who have a relatively fast catch-up process. Once a network starts to decline in some country, other countries tend to follow this process (diffusion synchronization). This explains why innovation followers have a much smaller and less pervasive network (islands of technology; isolated links) in the declining phase of networks than the early adopters, so, the nature of the technology leap differs between leaders and laggers. It also explains that once a country has missed this 'innovation window', it is unlikely that a certain transport system will capture more than a minor market share (low catch-up potential).

It is important to observe, however, that the use of certain technologies in an international or global network is more or less obligatory. Once certain technologies are used worldwide, no country can afford to refrain from their use. Containers and air transport are two major examples of this type of 'obligatory' technology. The use of other technologies is voluntary, which means that a country or region does not have to use a certain technology window, for example because the new technology does not fit in its techno-social structure, or because it is too expensive. Waiting for more acceptable alternatives is then an obvious strategic option for society (Rico, 1993).

4.2.3 Decline

The phase of decline shows another interesting phenomenon; physical infrastructure has a very long lifetime, covering usually more than one successive transport mode. For example, the beds of former canals were often used for new railway lines, which in turn were sometimes replaced by motorways and even long-distance cycle tracks and walkways; industrial harbours are now used for leisure shipping (e.g. yachts) and major parts of the motorway network have been built on ancient, unpaved trackways. In other words, the physical structure often remains, but its use or services will change.

Besides the long wave, there are also short waves which characterize changes within a product or a production process. In the automobile industry such waves cover a period of 10 to 30 years. Due to globalization, which also affects the transport industry, it is very likely that the short waves will be reduced significantly, so that the pace of innovation will increase.

4.2.4 Technological opportunities

In the field of transport demand, one may speak of a minimum, upwardly changeable, transport quality level to be fulfilled by transport modes in order to be successful in attracting customers. This has led to the development of 'dominant basic development concepts' (Bilderbeek *et al.*, 1993). Developing vehicle concepts, which differ significantly from these reference concepts in terms of price/quality, organizational and institutional aspects and other parameters, is risky. This explains why technological development usually follows 'known trajectories'. Only if substantial disadvantages occur in these trajectories, is it considered worth trying new ones.

As a result, improvements of current modes and design of new modes which use existing infrastructure tend to have far better chances of introduction and acceptance than modes which need new infrastructure (compare, for instance, Sections 6.2 and 6.3). Radical changes are therefore not very likely in the short and medium term. Evolutionary shifts seem to be more promising. However, many incremental changes may have the same impact as drastic changes.

In conclusion, if drastic changes are to be expected from technological breakthroughs in the transport sector, it will - in light of the life-cycle trajectories - take a long time before the sustainability criteria will be met. It may also be concluded that the introduction of new technologies is a continuous process. It is important, therefore, that in R & D sufficient attention is paid to the environmental impacts of new transportation technology. Often difficult trade-offs have to be made, since there may be conflicting objectives. For example, safety targets favour bigger and heavier cars, whereas curbing the greenhouse effect urges the introduction of smaller and lighter cars. In practice, marketing considerations also tend to be more important than environmental targets.

It should be recognized, however, that the introduction of new technologies is not an autonomous process. The outcome of this process is determined in areas outside the technology regime. In a way, technology is a response to challenges created in spatial, institutional, economic and socio-psychological fields (see also Rienstra *et al.*, 1996). All these aspects will now be discussed further in the remainder of this chapter.

4.3 SPATIAL FACTORS

This section will first briefly discuss spatial factors in relation to transport needs, which emerge when functionally dependent human activities are separated in space. Transport needs stem basically from three characteristics of the spatial organization of our society, namely:

- the level of spatial specialization (and interdependency) of human activities;

- the level of concentration of these activities;
- the distances involved.

Since transport is essentially an indirectly determined demand, the existing spatial pattern of activities is decisive for the need for spatial interaction and communication. This will briefly be elaborated in the next subsection.

4.3.1 Location patterns and transport

Existing locations (residences, industries, public services, recreational areas, etc.) determine the transport needs in the short run. Consequently, land-use planning, territorial planning or physical planning are important policy intervention measures in coping with transportation issues. Therefore, some remarks on land-use planning are in order here. There are some fundamental principles related to land use (Owens, 1992; Owens and Rickaby, 1992).

First, apart from land reclamation projects (usually in the sea), the quantity of space and land is limited. The use of land can, however, to some extent be intensified by using the 'third dimension', the air and subterranean space.

Secondly, various types of land use are not compatible. For example, negative external effects may exclude activities in close proximity to each other; another possibility is that a high market price for certain specific categories of land makes land use with a low return on investment impossible.

And thirdly, land use is significantly influenced by institutional measures imposed in spatial planning.

In this respect European countries have different traditions. For example, the Netherlands and the UK have a relatively well-developed planning system on various spatial scales, whereas countries such as Italy and Greece have systems that provide scope for many degrees of freedom. Furthermore, the role of spatial planning is generally limited when the aim is to change or reverse ongoing trends. The following circumstances are important in this respect:

- there is a large inertia in the built environment, particularly during periods of low economic growth (see Section 4.3.2);
- there is a large institutional inertia among policy makers (planners) which prevents them flexibly adopting new frameworks of reference for alternative locational patterns;
- there is a large variety of often mutually contradictory aims of spatial planning, and spatial policy is largely ineffective in reaching these aims.

Regarding the objectives of spatial planning for a reduction of transport needs, it should be emphasized that there is still a lack of knowledge about the principles underlying policy making for residential and work areas (Banister and Watson, 1994; Wood *et al.*, 1994). Much research has focused on the relationship between urban form and passenger transport. Urban form in this context means size and density, and one of the major conclusions so far is that higher density cities are associated with a high use of public transport and with a low gasoline consumption (Newman and Kenworthy, 1989). What matters, however, is where the interdependent work places and dwellings are located within the metropolitan area; in other words, the socio-economic composition of the city is a crucial element (Banister and Watson, 1994; ECMT/OECD, 1995; Wegener, 1995). In this way, two extreme spatial models, viz. a compact and a decentralized city, may be distinguished (see also Section 3.5.4).

4.3.2 Spatial success and failure factors

We now explore the role of spatial success and failure factors in achieving a sustainable transport system. Clearly, success and failure is co-determined by the relevant stages of a transport system's life cycle. But also the spatial embeddedness of transport systems plays a crucial role. Three broad categories of spatial success and failure factors will be discussed:

- spatial threshold and critical level factors;
- local and regional impacts;
- spatial inertia of infrastructure networks.

Spatial thresholds and critical levels

Spatial threshold factors are concerned with the minimum volume of passengers between given points, that are necessary for a transportation mode to be in operation and economically feasible. Such threshold factors usually apply to collective transport modes. In this respect, barriers to adoption arise when spatial threshold (minimum) levels of demand for the advocated technologies are not reached, for example, due to a low population density. Upper level spatial threshold factors are different, in that they are associated with a particular type of vehicle and the maximum distance it can bridge. Barriers may arise when the transport distance needed exceeds the critical upper level of the spatial range of the vehicle in question. In this section we will consider these levels in particular with regard to urbanization on the (inter)national scale and on the scale of metropolitan areas.

At the international level, large-scale patterns of urbanization are emerging leading to particular types of transportation needs (Masser *et al.*, 1992 and 1993; Nijkamp, 1993). In view of this, the specialization and concentration versus the chains and zones model (see Section 3.5.3) may lead to entirely different transport systems. The concentration pattern is geared towards radial transportation systems, whereas the chains and zones pattern is geared towards criss-cross transport relations.

Local and regional impacts

As mentioned above, various types of land use are mutually incompatible, meaning that they cannot be undertaken adequately in close proximity. Although we see in practice many variations between nations, currently a common understanding is growing about what is compatible and what is not. This is also reflected in (inter)nationally accepted maximum levels of exposure to various forms of danger, based on insights mainly into severe locally concentrated, negative external effects of certain types of land use (e.g. explosive or poisonous materials). The most common of these external effects for transport have already been discussed in Chapter 2.

One spatial factor should be added in this framework, which follows from the market value of land. Under certain conditions, the value of land has increased so much (as in inner-city areas) that it only permits those types of land use which generate a high return on investment, while excluding others. Examples are the CBD developments in American or Japanese cities.

Impacts of new transport systems at a higher geographical scale in Europe are often discussed within the framework of high-speed links between cities and city-systems. In this respect, a major issue is whether high-speed links increase disparities between economic core areas and peripheral (non-connected) regions, or raise the level of economic activities in such a way that both types of regions can benefit. Apart from the direction of the impact, there is the issue of causality mechanisms between transport infrastructure and regional development (Rietveld and Nijkamp, 1993; Vickerman, 1991).

Spatial inertia of infrastructure networks

A powerful spatial barrier to adopting a new transport technology is spatial inertia of the built environment and of infrastructure networks. Built forms following from land use, such as housing blocks, industrial estates and transport infrastructure, have a long life cycle related to the capital investment involved. As a result, various types of land use are fixed for a number of decades. So, once the infrastructure is built, it will be there for a long period (especially in historical city areas). As a result, technologies which involve step-by-step (incremental) or small change may have a better chance of adoption than technologies requiring radical change of infrastructure and land use. As shown above, this also happens because of technical inertia.

4.4 INSTITUTIONAL FACTORS

It is increasingly recognized that institutional factors may be of major importance for the successful introduction of new technologies (Opschoor, 1994). The main institutional agency is the government which has a far-reaching and regulatory influence on the transport market. Therefore, we will first analyse the role of the government. Next, we will discuss which barriers and opportunities may be distinguished in the institutional field.

4.4.1 Public policy and transport

Reasons for government interference in the transport market

The influence of the government on the transport sector is in most countries very strong, as there is an abundance of regulations, taxes and so on. There are several reasons for government interference in the transport market (Fokkema and Nijkamp, 1994; Nijkamp and Rienstra, 1995):

- positive external effects of infrastructure and of the transport sector in general, which favour the competiveness of the national economy and the generation of economic growth; this is the reason why a high-quality infrastructure is considered to be of strategic national importance;
- negative external effects (see also Chapter 2), which are not taken into account in the market decisions of an individual actor, such as air pollution or noise nuisance;
- several specific characteristics of investments in infrastructure, which make them less interesting for the private sector, such as lumpiness and indivisibility of large-scale investments (see Section 4.5);
- increasing returns to scale (up to a certain level of demand), which lead to a natural monopoly and give the possibility of abusing market power;
- infrastructure as a semi-public good, which means that exclusion from use is possible but often not accepted, because everyone is supposed to have a right to at least some level of mobility. This is the equity argument used in transport policy.

From the previous observations it follows that governments should be involved in decisions about the construction and operation of modes, in order to achieve a socially optimal level of investment, which means that the external effects should also be taken into account. It is, for example, very possible that a new transport mode has positive - or less negative - external effects compared to existing modes. Therefore it may be socially optimal that the government supports this mode, for example, by funding, but also by means of restricting regulations or taxes for other modes in order to internalize these external effects. Then it is possible to achieve an environmentally (more) optimal modal split.

But even when the government is not financing and exploiting the infrastructure, the government still has an important role because of the necessary planning and decision-making procedures required to build the infrastructure. It is also important that the government plans (and finances) complementary connecting infrastructure and the infrastructure for the transport to and from the nodes.

The design and execution of a proper transport policy is crucial, because it determines the competitive relations between the transport modes and the possibilities for certain transport modes to increase their market share. The construction of infrastructure has to be approved by the government, while the price-setting for the different modes was also until recently regulated by the government in almost all European countries. This is why the profitability and the level of competition is essentially determined by public policy.

Therefore, it is clear that it is impossible to develop, introduce and exploit new transport modes without support from the government. The new mode has to fit in the existing transport policy, so that the government can ensure sufficient support for a project. It is also important that the government guarantees not to change the policy for the mode concerned in an unexpected negative way (in order to avoid political risks of investments).

Decision taking by the government

Many of the possible barriers to the introduction of new technological options are caused by the 'peculiarities' of public decision making (Frey, 1983; Rietveld, 1997). The development of policy and the decision making by the government differs from that of the other players in the market. Social interests are particularly important here. An example is the protection of the country's national transport industry (e.g. subsidizing the purchase of new cars in France and Spain); another is the stimulation of the export of a transport system technology (e.g. the Maglev high-speed train in Germany).

Characteristics of the decision making by the government are that there is a trade-off of several - often conflicting - interests; the process is highly formalized and structured via formal procedures, with formal powers strongly divided between the various government bodies (both horizontal and vertical). In addition to this formal power of agents, in policy networks informal power (above all among private players) also exerts an important influence. The informal power of an agent is partly determined by:

- the degree to which other agents are dependent on other agents;
- the knowledge and information available to an agent;
- power and dependence relations that are the result of historical factors.

Partly as a result of the distribution of formal and informal power among the agents (in policy networks), the influence of conflict situations on the decision making (e.g. delays) may be considerable. Conflict situations can occur if there are external effects and distributive effects as a result of the project concerned. Most clearly present are conflicts of interest in large infrastructure projects: strong local negative environmental and ecological impacts versus economic and environmental benefits on a larger (i.e. national or international) scale.

The trend towards deregulation and privatization may mean that the government adopts a less active attitude and leaves the realization of policy to participants in the market. The implication of this is not always clear, as it may have both a retarding and an accelerating effect; retarding, since more parties are required to realize a project and conflicting interests may increase (coordination problem), acceleration, because market orientation and efficiency may be expected to increase.

4.4.2 Institutional success and failure factors

Various success and failure factors may occur in every technological innovation or introduction of distinct transport systems. Barriers that are fairly general in nature are, for instance, problems related to the need for space, environmental nuisance or financing problems. Obvious questions are then: Who is going to invest? Is there a basis for consensus? Who will bear the costs? The claims made on space, environment and (government) funds play an important role in the political decision making process. It should be noted that institutional processes are not autonomous, but also follow the life trajectories of transport systems, they may play a conditioning or proactive role in changing or initiating such trajectories.

Innovations in the transport system inevitably seem to invoke barriers of a varied nature. However, innovations may also exert a positive effect. Advanced transport systems may develop, which create a certain new image for the area concerned. Innovations may also increase the competitiveness of a transport system, and as a result generate economic and social benefits.

At the same time, often there is also some hesitation about introducing new technologies, because of the risks involved. The costs of starting and malfunctioning are preferably left to others.

With regard to technological changes in transport systems a distinction may be made between the adaptation of an existing transport system and the development of a new one. The same holds for collective transport systems in comparison to individual transport systems, or national/international transport systems versus intraurban and interurban transport systems. These categories will be discussed below.

New transport systems versus improvement of existing transport systems

The viability of a transport system depends to a large extent on the return on investments. Innovation in an existing transport system has the advantage that the existing facilities can be employed as the basis for a start. As a result, the additional investments needed for the innovation are relatively low (in comparison to investments already made). New transport systems have either to compete with existing ones or develop a new demand/market niche, which may be much harder.

With the introduction of new transport systems much more adaptation of legislation and regulations is also necessary in many cases. In addition, new transport systems often take up more space than is needed for the expansion and improvement of existing systems, since the latter may partly use existing infrastructure and have already put a considerable stress on scarce space.

Collective transport systems versus individual transport systems

Collective transport systems, as a result of their high transport capacity, function optimally on routes with a high spatially concentrated transport demand (see Section 3.5).

A second way in which collective systems may be distinguished from individual ones concerns their dependence on supplementary transport systems. Travelling with collective modes is intermodal by nature, while individual modes offer door-to-door transport. This makes the functioning of collective modes dependent on their connectivity with other transport systems that offer transport to and from the nodes. Coordination problems between different modes may be an important failure factor in this case.

Individual and collective modes differ also with regard to the level of government involvement in the operation of the system. For collective modes government intervention is much higher, since these are mostly highly subsidized and mostly publicly-owned. Therefore, collective modes depend more on public policy than individual modes.

International transport systems versus urban transport systems

International and national transport systems cross numerous administrative boundaries (national, regional and municipal). As a result, horizontal and vertical coordination problems occur frequently in the creation of such systems. Regarding transnational networks there is no single authority with full responsibility, but planning and decision making takes place via negotiations. This results in some failure factors:

- projects are no longer viewed solely on their individual merits, but become part of a total package of projects which are negotiated between the countries concerned;
- standardization problems occur (horizontal coordination between governments), e.g. the development of high-speed trains in Europe has for decades been a matter of national interest (Germany, France, Spain, Italy, the UK). As a result, each country has developed its own high-speed train technology.

A further coordination problem relates to the tendency of national governments to upgrade their existing and new networks according to national demands, instead of European ones (Nijkamp and Vleugel, 1993). This is due to the fact that the financing and decision making are the responsibility of national governments. Expansion and better utilization of present infrastructure of international importance must also be placed in an international framework. It is noteworthy in this context that current discussions on better utilization of the present rail network are, above all, nationally oriented, while rail transport itself is highly competitive at longer distances.

At the national level constitutional factors also play an important role. In centralist administrations the national government is less bound by cooperation from lower authorities. The means of power that ensures rapid progress in France is the *Déclaration d'Utilité Publique*, a kind of emergency law that the French Council of State can trigger if national interests are at stake. In countries where local autonomy is greater, one sees a tendency to develop similar powers at the national level to prevent delays as a result of possible vertical and horizontal coordination problems (NIMBY legislation).

For urban transport systems lack of space and environmental aspects are more important. The infrastructure has to be built almost entirely in intensively urbanized territories, which may increase social resistance. A second barrier is the availability of funds, which may be more limited at lower scale levels.

In conclusion, a significant part of the innovative ability of transport policy is related to institutional and legislative factors.

4.5 ECONOMIC FACTORS

In the development and implementation of new transport modes four stages may be distinguished, which are important for the economic success and failure of transport modes. These stages can be analysed from the viewpoint of a private investor, because this is the best way to understand underlying economic principles and factors. The first is the *R & D stage*, followed by the *financing stage*, the *construction stage* and finally the *exploitation stage*. Economic factors may be a driving force for the introduction of new transport systems, but economic development may also be triggered by transport investments (Bruinsma, 1994), so that there is essentially a mutual linkage in the time paths of economic aspects and systemic aspects of a transport systems' evolution.

All these stages are important for the economic success of transport modes. In the R & D stage the new mode is developed, investigated and tested. It is important that economic factors like the market potential are also be considered. In addition, political decisions have to be made about the introduction of the new mode. The outcomes of this stage will be taken as given in the remainder of this section. Most attention will be paid to the financing stage, because this is when the economic possibilities have to be assessed. Next, a short analysis of past privately financed projects will be presented, and finally relevant success and failure factors are summarized.

4.5.1 The financing stage

Most investment costs of a project are normally made when the infrastructure is constructed; the other costs (e.g. rolling stock) are a smaller part of total costs. According to Emanuel (1991), 80 per cent of the costs of a newly realized transport mode are construction costs. For society as a whole, the infrastructure costs are even more important, because the external costs of the new transport mode are mainly discussed during the decision procedures about the construction and location of the infrastructure.

Since the beginning of the century most infrastructure has been financed by the government or by organizations which are tied to the government. During the last few years, however, more emphasis has been put on the participation of the private sector in financing and operating the infrastructure. The exploitation of the infrastructure was, until recently, the task of public companies. However, in many countries these companies have become increasingly independent of government influence or are even privatized. This puts more emphasis on the economic return of investments in infrastructure, while the importance of the social return receives less attention.

By analysing the economic success and failure factors at this stage, five important issues may be distinguished. These are (1) the characteristics of investments in infrastructure; (2) the risks; (3) the expected costs; (4) the expected revenues; and (5) the required characteristics of the new transport mode or technology.

Characteristics of investments in infrastructure

Investments in infrastructure may be distinguished from competing investments such as immovable and capital goods in several ways. Characteristics of investments in infrastructure are (ECMT, 1990):

- The expectation of the economic life of an infrastructure is very long compared to other investments. This may vary from 20 years to more than a century. The payback period of infrastructure investments is also long; concessions are often granted by the government for 15 to 30 years. Other immovable and capital goods usually have a much shorter payback period, the average for capital goods is 8 to 9 years.
- The operational costs are relatively low, especially for the longer distances. There are some labour, overhead and maintenance costs, while a system must be set up for selling tickets and making reservations. Compared to the

construction costs of the infrastructure and to the exploitation costs of other investments, these costs are low, however.
- At the beginning of the project a large amount of capital is required. Often also high loans have to be attracted, which make interest costs high.
- The period before construction actually begins can be very long, because of many legal decision-making procedures, resistance in society, etc.
- The construction period is very long compared to other investments; this period may take 2 to 7 years - in this period there are no revenues, while there are already capital interest and other costs being incurred.
- Once started, the investment is almost irreversible - if the construction is stopped, this would imply total loss of the capital, because it is not possible to use the investment elsewhere.

From this it follows that in the early days of a project much capital is needed. In general, this makes the interest costs very high at the start of the project, when the cash flow and the return on investments are low. In most cases there are no revenues at all before the exploitation starts. When it starts, the profits get higher over time, because more repayments are made, which reduces the interest costs. When mobility grows, the revenues are growing too. The problem is that these high profits and revenues often start to flow decades after the investment, which makes the uncertainty and the risks very high.

In practice, however, it is quite possible that there is no profit at all (Nijkamp and Rienstra, 1995). The construction costs of infrastructure are (up to a certain level of demand/transport) fixed costs; the other costs are partly fixed and partly variable. From this it follows that compared to competing investments fixed costs are very high for an investor, while variable and marginal costs are very low. When the price in this case is set according to the marginal costs, it is often not possible to make a satisfactory return on investments, thus precluding private sector investments.

Risks of investments in transport modes

Investments in infrastructure, like all investments, involve risks, which may be distinguished as follows (OECD, 1987b; McKay, 1989):

- **political** risks; for example, changes in the transport policy, nationalization, intervention/regulation by the government, etc.;
- **financial** risks; for example, fluctuations in interest rates, fluctuations in exchange rates, wrong expectations about inflation, etc.;
- **construction** risks; for example, delays, unexpected and higher costs (especially when new technologies are used), etc.;
- **operational** risks; for example, damage by accidents, vandalism, etc.;

- **commercial** risks; for example, wrong cost estimates, wrong estimates of the traffic volume, unexpected competition, etc.

Most of these risks are included in all kinds of investments, but for investments in infrastructure they are particularly high. This is because of the long payback period; therefore, estimates have to be made for a very long period. When a new transport mode is financed, the construction and commercial risks are extremely high, because new and unfamiliar technologies are used and the market potential is unknown.

The political risks, however, are the most important difference with respect to other investments. In practice, governments always wish to influence the planning of infrastructure, because of the important positive and negative external effects and the national importance of high-quality infrastructure (see Section 4.4.1).

There is always also a danger of changes in laws or of new regulations or even of nationalization, because of a changing transport policy. For example, in the Netherlands there is jurisprudence on this subject, which states that the government is not bound to contracts if national interests are harmed by unforeseen circumstances.

In conclusion, the risks of investments in new transport modes are very high compared to other investment opportunities; this in turn makes these investments unattractive for private investors. Alternatively, a high risk compensation is needed for the private sector. In theory, this compensation must emerge out of high profit expectations, as is also shown by recent tunnel projects in the Netherlands. Another possibility, however, is that the government guarantees (part of) the revenues in one way or another, or makes the investment attractive in some way (e.g. via tax exemptions).

The expected investment costs of transport modes

Important for the calculation of the return on investments are the expected costs and revenues. Nevertheless, it appears to be very difficult to estimate the costs of construction of major infrastructure projects. These projects often turn out to be much more expensive than was originally estimated; well-known examples are the Channel tunnel, high-speed train tracks and the Betuwe freight railway line in the Netherlands. This problem arises especially when the project is a completely new transport mode or when new technologies are used. Then many costs are not known at the beginning of the project and the original estimates are too low in almost every case.

Another important cause of cost over-runs are the more expensive solutions, chosen because of resistance in society against environmental decay caused by new infrastructure. Thus often additional costs have to be made to avoid external effects (leading, for example, to (half-) subterranean infrastructure and noise shields). It is, however, very important that the cost estimates are made in a reliable and accountable way, as otherwise it is impossible to make a conclusive statement about the economic viability of the project, because no proper return on investment can be calculated.

The expected revenues of investments in transport modes

For the assessment of the expected revenues of a project several issues are important. The main task is a proper analysis of the market in which the mode will operate. First, it is important to analyse the (sub)market(s) the new mode is aiming at (Gunn *et al.*, 1992). These submarkets can be distinguished according to the residence of clients, their destinations, the reasons for travelling, etc. It is, for example, essential to know which segment of the market the project is aiming at. When it is a mass transport mode, the price must be low, while comfort, speed and service are of lesser importance for most people. When the project is targeted towards the higher level business market, speed, reliability, service and comfort are very important, while the price may be set somewhat higher. The latter is, of course, also dependent on the prices of competing modes.

It is also essential to consider the changes that are expected in the submarket(s) in the future. Several issues regarding the future of passenger transportation may be distinguished (Nijkamp *et al.*, 1994):

- **demographic** factors; for example, declining population density in urban areas, the age structure, migration, labour participation, decline in working hours, etc.;
- **political** factors; for example, European integration, the opening of new markets in Eastern Europe, the transport and environmental policy of the government and/or municipalities, etc.;
- **socio-economic** factors; for example, economic growth, growth of and changes in trade flows, spread of production, the development of trade blocks, etc.;
- **technical** factors; for example, the arrival of new competing transport modes, improvements of existing modes and telecommunications, etc.

After this analysis the number of travellers who will use the mode can be estimated. Two groups are important here: first, the travellers who will use the new mode instead of an existing one (substitution); and second, the group who did not travel on the route before (generation). It is then important to investigate how great the uncertainty is about these estimates. After this the optimal price can be set; eventually there may be price differentiation in one way or another, to reach different submarkets.

For creating total revenues it may also be important to generate revenues other than those which are directly related to transportation as such. Catering on long-distance journeys, shopping facilities like tax-free shopping at airports are well-known examples of indirect revenues. But it is also possible to think of other possibilities to generate revenues. For example, stations are popular location sites for several public-oriented kiosks, snackbars, florists, travel agencies, etc. Another possibility for generating revenues is the value of the sites around railway stations, as their prices usually rise after the building of a station. The question here is, of course, whether it is possible for the investor in the project to capitalize on such external benefits.

When all these revenues are known, the expected revenues for the investor may be assessed for the long run.

Required characteristics of the new transport mode

The characteristics of the new mode are critical for attracting travellers. So it is necessary to know what factors new clients find most attractive (Emanuel, 1991). Several factors may be distinguished here. In the first place the new mode should have sufficient capacity, including in rush-hours. The service of the new mode has to be satisfactory; in particular the waiting times should be short; thus a high frequency of vehicles is needed, in off-peak hours, too. The travelling time should also be acceptable.

Also vital is the transport to and from the nodes and the location of these nodes, especially for collective modes. This appears to be one of the main reasons why travellers choose a private transport mode and is one of the main explanations for the success of the private car, which is not dependent on other modes, but offers door-to-door transport itself.

Another important characteristic of new transport modes is their external effects, which may be favourable compared to existing transport modes. It is therefore imperative that the society and the government support the construction of these environmentally friendly modes. From this it follows that the negative external effects must be kept as small as possible.

4.5.2 The construction and exploitation stage

When the financiers of a project (private and/or public) are found and the procedures completed, the construction of the project can start. During this stage the costs must be kept under control, because this often causes a lot of problems (e.g. the Channel tunnel). It appears that often the costs of construction are much higher than expected, for instance, because expensive items have been forgotten in the estimates.

It is also important that the construction contracts are clear about who pays over-runs; this often leads to conflicts, which can also have economic implications (e.g. delays, higher costs).

Reducing disturbance during the construction is essential, especially when the construction takes place in densely populated areas. When the nuisance level is high, the resistance in society against the construction will be strong, which may cause additional delays and hence higher costs.

A good marketing strategy is needed to get a profitable market share. It is common practice for ticket prices to be quite low in the beginning, to attract new travellers and to give travellers using an existing mode an incentive to change transport modes. Later the charge can be raised to the economically (or socially) optimal level.

4.5.3 Lessons from existing projects

From reviewing existing practices on private initiatives in the transport sector, it also seems very important that the traffic flows are substantial (Rienstra and Nijkamp, 1997b). For example, projects which reduce the impact of barriers (borders and natural barriers) are attractive, since these may serve a large market. As seen before, the fixed costs of projects are often high and the variable costs low. So a project is in most cases economically successful when the fixed costs can be shared by many travellers. The price can then be kept low, which makes the mode more attractive.

4.5.4 Economic success and failure factors

From the analysis in this section, it appears that when a new mode is technically feasible, a series of economic factors appears to be fundamental in explaining the success and failure factors of new transport modes.

First, the (fixed) investment *costs of infrastructure* appear to be a major barrier. The characteristics of these investments make these very risky for the private sector. This holds especially for the construction of infrastructure for new modes, which have not proven to be successful. In this respect, it is often the task for governments to stimulate these investments by making them more attractive.

Next, the (variable) *costs of exploitation* are relevant. This depends largely on the type of mode. Such costs may be the collection of fees, information provision to users, service provisions, maintenance costs, etc. These costs also depend largely on the way the infrastructure is financed.

The direct revenues of a new mode of transportation are determined by the price per passenger and the number of passengers. So the *tariffs* of a new mode depend on the market segment the mode has to serve, e.g. cheap mass transit or high-speed luxurious transport. To cover several submarkets, price differentiation might also be necessary. Because of the high fixed investment costs and the low marginal costs also the *seat occupancy rate* is important. This depends very much on the size of the market and on the level of the peak demand. When the latter is high, the average seat occupancy rate is low, which results in unprofitable capacity.

Besides direct revenues, the generation of *indirect revenues* may also be required, in order to reach an acceptable return on the investment. These revenues depend on the type of mode.

To attract investors, the *risks* have to be reduced as much as possible. These risks result especially from the long payback period and the high political risks. To reduce the risks may be partly a task of governments, although part of these risks also may be regarded as a private risks for which a risk premium is granted.

So it may be concluded that new technologies face several economic success factors, and that investments in these technologies often may not always be attractive, in particular for private financiers.

4.6 SOCIO-PSYCHOLOGICAL FACTORS

Several socio-psychological factors may influence the success of new transport technologies. This holds especially when a new system does not adequately meet existing needs, perceptions and standards of the target group concerning travel behaviour and travel mode use. In that case, an insufficient social basis may exert an important restraining influence on the process of acceptance and adoption.

A recent example is the shuttle-bus project in the Dutch city of Groningen, where fast bus services in the morning and evening peak between neighbouring commuter municipalities and Groningen city were offered, with a guaranteed seat and a morning newspaper for the season ticket holder. The project, which was mainly meant to persuade car commuters to travel by bus instead of by car, was discontinued after an experimental period of more than two years, because of lack of interest. What is remarkable is that an interim user survey indicated that a majority of the shuttle-bus travellers had been regular public transport users in the past rather than car users.

Similar shuttle-bus projects elsewhere in the Netherlands (especially in the Randstad) have turned out to be relatively successful, however. The untimely termination of the project in Groningen may be attributed to the fact that the congestion problem in the north of the country is not as severe as in the Randstad. Traffic jams are (as yet) very rare in the north, which apparently reduced the incentive to change the travel mode.

To identify which factors are most important, we will discuss several strategic key forces. First, 'free-rider behaviour' is discussed, followed by the relevant (socio-psychological) determinants of travel behaviour and mode use. As stated before, these forces are particularly significant in generating the basis for the acceptance and adoption of technologies, an issue which will be discussed next.

Our discussion is inevitably rather broad, particularly because there is no abundant research on psychological success and failure factors of possible future travel modes and systems. Moreover, often external social factors are of importance - like the socio-economic situation - which are difficult to specify in the long term.

4.6.1 Free rider behaviour

From the above-mentioned Groningen case it may be concluded that there is a relatively low willingness among people to change their travel behaviour and travel mode choice with the aim of anticipating problems which are as yet not clearly perceptible and manifest. This may be closely connected with the social dilemma nature of the mobility problem. This means that, in connection with a behavioural choice, there is an actual or assumed discrepancy between individual interests (mostly short term) and public interests which are of importance in the long run (Liebrand and Van Lange, 1989). It appears that people mostly let their individual interests prevail, thus at the expense of public interests (free rider behaviour). This is especially the case when negative social effects only become evident in the long run (Olson, 1965; Schmidt, 1989). Michon's view (1980) that human behaviour is highly 'failure driven' also fits in the above-mentioned tendency. This means that as long as a certain behaviour pattern is satisfactory for someone, the motivation to change or adapt his or her behaviour will be small. Only when negative effects become predominant, are people inclined to reconsider and if possible change their behaviour.

As a result, for example, nobody (of the final users) would object to local buses running on natural gas instead of on diesel oil. At the same time nobody would make more or less use of the bus, just because it now runs on natural gas. This makes it interesting to analyse which factors influence travel behaviour and travel mode choice.

4.6.2 Determinants of travel behaviour and travel mode use

Mobility patterns are generally considered to be a derivative of other activities, which are considered necessary and/or desired and which a person cannot or does not want to execute at home (Van der Hoorn, 1989). Thus the most important determinant of travel behaviour is immediately evident: the travel need, which means the necessity or wish to travel. This need can be determined by situational circumstances (e.g. business travel) as well as intrinsic motivation (e.g. socio-recreational travel), which are closely linked to the activity pattern of the individual in time and space. The travel need is likely to change when:

- a significant change in the daily activity pattern occurs (leaving the parental home, having a (new) job, becoming unemployed, commencing teleworking);
- the travel possibilities change, which concerns not only the availability of travel modes but also of infrastructure. In addition, this holds for subjective travel possibilities, i.e. the possibilities as they are perceived by the person. These may differ in nature and number from the objective situation.

As is presented in Figure 4.1, travel needs and possibilities are not independent, but strongly influence each other (De Boer, 1989).

Possibilities are determined by and tuned in to the travel need, while the supply of new possibilities often generates new demand, as has been shown in the past. A corresponding trend can be observed at the micro-level; purchasing a car in a household often means more use of the car than was presumed before buying it (Onnen and Van Knippenberg, 1989). At the same time, an elimination of the travel possibilities - e.g. because of selling a car or closing public infrastructure - often appears to result in a decrease in travel demand.

The travel mode choice is first of all determined by the individual travel need and the (perceived) travel possibilities. The final choice from the possible travel modes is made on the basis of a cognitive and motivational decision-making process, making use of the available knowledge of the alternatives, perceived or presupposed pros and cons, personal preferences and attitudes and social and individual norms. In other words, the alternatives are balanced against each other, which is, however, only partly rational and very strongly biased (see Rooijers and Steg, 1991). The above-mentioned factors will be discussed briefly, partly using the behavioural theories of Fishbein and Ajzen (1975) and Ronis *et al.* (1989).

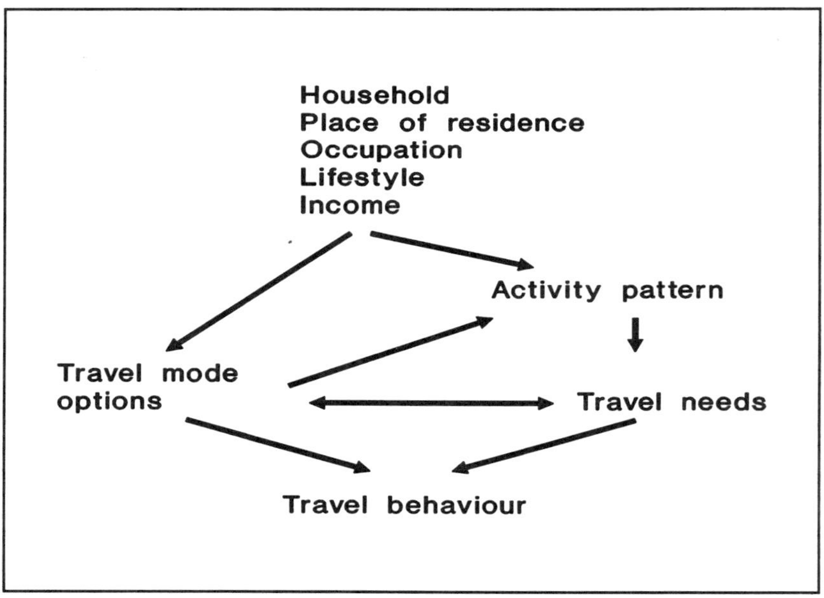

Figure 4.1. Model of travel behaviour

Knowledge

An optimal rational choice from the possible travel modes requires that all objective possibilities are known. As mentioned before, this is often not the case. Particularly those people who travel little or not at all by public transport hardly see it as an alternative. The knowledge from which a choice has to be made is usually just a subset of all possibilities; this leads to the first subjective bias in the decision-making process.

Perceived pros and cons

An optimal objective choice further presumes that all pros and cons of the alternatives are known, and these then have to be weighted on a rational basis. In reality, these conditions are not fulfilled, because they are both largely biased. As a result of selective perception and cognitive dissonance mechanisms (Festinger, 1954), people in general are inclined to overestimate the advantages and to underestimate the disadvantages of their own regular way of transport, while for the alternatives the opposite holds true. For example, the costs, travelling time and environmental pollution linked with car use are often trivialized, whereas long travelling times and lack of comfort in public transport and bicycles are emphasized (Rooijers, 1992b). However, it should be noted that the characteristics mentioned as being the most important by practically all traveller groups (public transport users included) are particularly associated with the car (short travelling time, comfort, flexibility, easy reach of destinations).

Individual preferences and attitudes

Besides these cognitive factors, motivational and even emotional factors play a significant role. Such factors include aspects such as pleasure, privacy, status, personal control and representativeness, which contribute greatly to the preference for individual travel modes. This motivational and emotional preference is more apparent among men than women, because women tend to choose their travel mode more because of functional considerations (Bureau Ferro, 1991).

Social and personal values

People do not live in isolation, but act in a social context, which may have a substantial influence on the decision-making process and the behaviour of a person. In fact, people are strongly motivated to compare themselves with other relevant groups and to conform to general trends in the social context (Festinger, 1957). Here, too, it should be observed especially that individually perceived social norms are important, which differ from objective standards. So the result is that the perceived social norms are distorted or influenced by individual preferences and attitudes.

Not only social but personal norms may be relevant as well. These are the individual's behavioural motives, which are of importance for him- or herself, but which are more or less isolated from individual preferences and perceived social norms. For instance, an individual may have a strong preference for car use, but nevertheless choose to use a bike for health reasons.

All these factors may play a role in the process of travel mode choice, but in what way the various cognitive, motivational and normative aspects are integrated is mostly not clear. In any case, the outcome of the decision-making process is that one of the alternatives is chosen and used. This leads to the conceptual model as presented in Figure 4.2.

However, the assumption is that the selection process is mainly limited to the travel mode selection regarding new and incidental travel. When it also takes into account repeated similar trips, positive experiences (or not having negative experiences) with a certain travel mode can act as a positive reinforcement of the choice made before, which increases the chance that on the next occasion the same mode will be chosen (Rooijers and Steg, 1991).

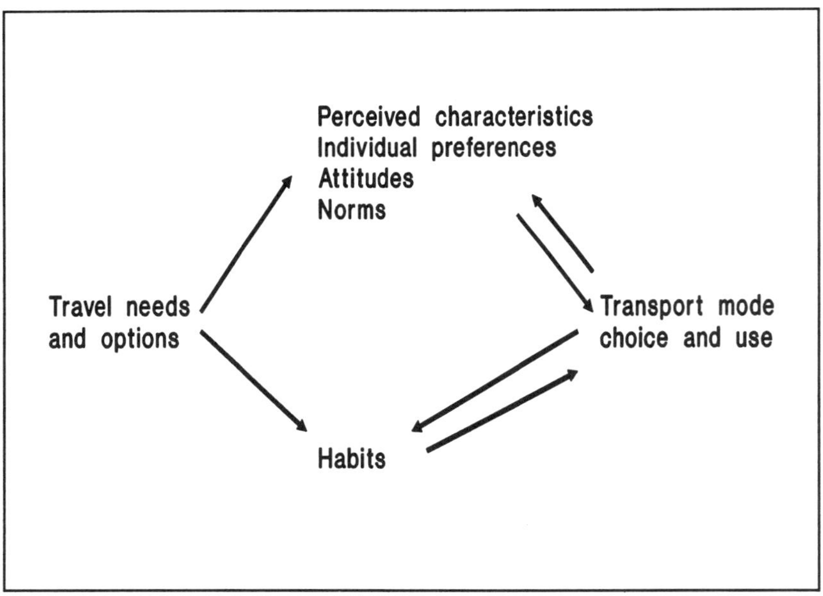

Figure 4.2. Model of travel mode choice and use

This may have at least two important consequences:

- there is hardly an explicit choice; the travel mode is used because it was used before, and therefore other alternatives are not even considered;

- persistent prejudices and misperceptions may develop which can hardly be remedied.

So the existence of habits hampers the possibility to influence behaviour. Here it should be observed that the chance of habit formation is probably the highest for car use, because the car is more multifunctional than other travel modes; therefore, it has a highly attractive value. As a result, car use will not quickly lead to negative experiences.

To break out of undesirable habitual behaviour the individual will first need to have knowledge about all relevant alternatives (see Ronis *et al.*, 1989). Subsequently the decision-making process itself should be influenced by control measures that affect the factors mentioned before.

4.6.3 Acceptance and adoption of technologies

In order to comprehend and - if possible - predict the acceptance and adoption of new transportation systems, it is important to understand the factors underlying mobility and travel mode use. After all, the introduction of new systems always occurs in relation to existing travel modes; so there is nearly always a matter of competition.

It should be noted that acceptance is not the same as adoption. Acceptance of a particular transport system means that there is a positive or impartial attitude towards (the introduction of) the system. Adoption means that the system is actively used for the desired mobility. It is perfectly possible for a particular new travel option to be accepted, but hardly or never used. Generally speaking, it can be said that acceptance is a necessary, but not sufficient, condition for adoption.

Although it is not always possible to draw a straight demarcation line, a rough distinction can be made between, on the one hand, adjustments in existing travel modes (like cars with a higher energy efficiency) and, on the one hand, the development of new systems, which implicitly or explicitly aim to decrease the use of current systems (like the development of the high-speed train). Apart from this, a distinction can be made between collective and individual modes. The resulting four subcategories will now briefly be discussed.

4.6.4 Improvement of current collective transport systems

Acceptance

It is clear that social opposition will be the lowest (and therefore acceptance the highest) for technological improvements of current collective transport systems. For example, small-scale experiments have been carried out with buses run on natural gas or biogas/bioalcohol. When people know this technology, travellers will be either mainly positive (usually from an environment point of view) or impartial to this.

Most existing collective transport systems are already electric-driven (tram, underground, train). Technological improvements which limit the pressure on environment (improving electricity generation, or interception and washing of fumes), may improve the energy efficiency. The traveller will not, or hardly, notice these types of improvement, however, and therefore there will be no opposition.

The introduction of advanced transport telematics systems is likely to have a great impact on public transport use (for a review see Nijkamp *et al.*, 1996b). The quality of public transport is likely to be improved significantly once smart information systems, automatic ticket systems (smartcards), and real-time traffic guidance and safety systems are introduced more widely (see also Emmerink, 1996).

The factors which hold for road and rail travel, are also likely to be valid for air travel. Technological developments which lead to a restriction of the negative environmental impacts (higher energy efficiency and/or other fuel systems) are certainly expected to be valued positively. The introduction of new, advanced and standardized air traffic control systems will be a major step forward, not only in environmental terms, but also in terms of reduced travel time and increased traffic safety.

Generally speaking, it can be said that technological adjustments of current collective transport systems will, as noted, cause little opposition. Considering the growing social awareness towards the environment, such improvements will probably be widely appreciated. This general principle holds as long as other relevant values are not at stake. For example, if improved collective transport systems become more expensive as a result of the adjustments, acceptance might turn into opposition after all. This does not only apply to financial consequences; the same thing applies if the safety of travellers is affected.

Adoption

Technological adjustments of current collective transport systems are not expected to have any influence on the need and possibilities to travel. Moreover, the engine system or the drive technology does not play a role in weighing the pros and the cons. As a result, it is likely that no effects on the use of the systems will occur. This means that it is almost completely up to transport companies and governments to decide whether to adopt and implement these technologies.

4.6.5 Technological adjustment of current individual transport systems

Acceptance

Acceptance of technological improvements of current individual modes (in particular cars and motor cycles) is less obvious than that of collective systems. This is due to the fact that the final user of individual transport modes is in an entirely different position, because the decision to adopt lies mainly with him or her and not with others.

In conclusion, people are very sensitive to measures which will have negative results for them and which will restrict their (perceived or supposed) freedom of choice. As a result, at least three general principles can be formulated with regard to the acceptance of improvements of current individual modes:

- adjustments to parts of the vehicle which play hardly any or no role in the decision to purchase will meet little opposition;
- adjustments which, *ceteris paribus*, will have positive results with respect to vehicle features and are considered to be important (e.g. a higher fuel efficiency) will meet little opposition;
- the degree of acceptance correlates negatively with the degree of restriction of the (perceived) freedom of decision-making.

It should be noted that the (perceived) freedom of decision-making is restricted not only by issuing rules and regulations, but also by physical and financial measures.

Adoption

There are two kinds of technological adjustments. First, there are adjustments in the area of vehicle technology, which are carried out by the industry without any say for the final user. These adjustments are made on all new vehicles, and therefore there is no adoption problem here. Second - and this is the main point - there are new technologies which are brought on the market alongside the existing ones; therefore these are new alternatives.

An essential difference between collective and individual modes in this respect is that an individual mode has to be purchased, after which this mode will be used instead of collective modes. As a result, people are much more psychologically involved with individual modes, which makes acceptance more difficult (Stokes and Hallett, 1992; Vlek and Michon, 1992).

If the vehicle, fitted with all the latest technology, has evident advantages, anticipates existing individual preferences, and is appreciated by the social environment, people will probably adopt the new technology. When one or several of these conditions are not satisfied, the adoption chance depends on the relative importance of these factors. Therefore, in those cases the prediction is very difficult.

An example may clarify this. The question of whether a car driver - confronted with buying the next car - is likely to opt for an electric car first depends on the knowledge of its existence. Subsequently, the pros and cons of an electric car against those of a conventional one (such as cost, user-friendliness, range, the current infrastructure for charging batteries, environmental consequences) are weighed up. The individual will take into account his or her personal preferences (like the producer, type, image and driving features of the car) as well as the perceived social standards (views from and use of electric cars by his or her social environment). The final choice depends on the relative importance the person attaches to the various aspects.

4.6.6 New collective transport systems

Acceptance

As was the case with improvements of existing systems, the degree of acceptance will be very high for new collective systems. However, other values which are considered to be important for potential users must not be affected, and hindrance to their daily lives should not be restricted. The reason for this 'yes, unless' position is also the same: the final user or consumer does not carry any responsibility for the introduction and management.

Safety and nature conservation are examples of important values which can lead to a great deal of opposition to and direct action against transport projects. However, the social unrest and opposition is possibly even higher if large groups of people experience or fear a great deal of hindrance from the new system in their immediate living environment. This point would be particularly relevant where the implementation demands completely new infrastructure (e.g. HST tracks).

Adoption

There are three relevant questions to determine what the new system means for mobility possibilities:

- Is the system, considered objectively, a good alternative in addition to the existing ones?
- Will the system coexist with the existing systems, or will it replace them? If the system is introduced, the number of travel mode choices will be expanded, which can complicate travel mode choice. Moreover, in that case the new system has no market segment of its own, in contrast to system replacement; this will increase competition.
- Will the new system be perceived as a valuable alternative?

With regard to mobility needs it is important whether the system offers a better alternative for existing travel needs or creates new needs, otherwise a sufficient degree of adoption will not occur.

It is difficult to indicate which cognitive and motivational aspects are considered to be most crucial and what their correlation is at the time of implementation. In addition to the advantages and disadvantages of current travel modes, this also depends on socio-economic developments, the degree of social awareness and the influence of marketing activities.

4.6.7 New individual modes

Acceptance

The chance of extensive social opposition is highest for the implementation of new individual transport systems, especially when these systems are not complementary. The reason for this is that, especially with individual travel modes - which are usually the property of the user, there is a great deal of involvement, and the individuality and freedom of choice of existing systems is very much appreciated, and perhaps even considered as a social achievement. This issue is essential in the development of more or less 'collectivized' individual systems, such as 'guided cars', and with the development of systems, which offer fewer user possibilities as a result of infrastructural restrictions. Yet, in a broader perspective such systems may become important because they may present a (partial) solution for both the environmental and the congestion issues associated with conventional individual transport.

Adoption

For adoption again the process of the purchase decision is most important. In principle this process will be similar to that of adoption of adjusted current travel modes. As a result the features of the new system will make or break the adoption. Considering the large number of factors involved the outcome is unknown.

4.6.8 Socio-psychological success and failure factors

In general, it may be concluded that technical improvements will only be accepted when individual lives and the freedom of choice are not affected. This applies particularly to collective modes. The adoption of such collective technologies may cause problems, however, because in the main the private car is preferred as a travel mode.

Improvements of individual modes may be harder to introduce, since people are much more psychologically involved with this mode; and it is also often important for people's lifestyle, etc. Therefore, introducing alternatives may cause problems with acceptance as well as with adoption. In general, therefore, improvements will generally be hard to introduce too.

4.7 CONCLUSIONS

Technologies have their own life cycle of emergence (birth), growth (market acceptance), maturity (market saturation) and decline. Whether or not a technology will enter these stages is dependent on many success and failure factors, which may be found in either the spatial, institutional, economic or socio-psychological fields. Especially for new transport modes, it may be very hard to cope with these factors and introduce them successfully, especially because the private car - and plane - fulfil most success factors to a large extent. Nevertheless, new or improved - more environmentally friendly - modes should have advantages over these modes. As a result, it is mainly improvements of current modes, using (at least) existing infrastructure networks which may be successful, while radical changes are less likely. Given the long life cycle of transport technologies and the existence of many potential impediments for introducing radically new systems, it seems likely that sustainable mobility will not easily be achieved in the short or medium term.

After this general analysis of success and failure factors, we will apply them to various transportation technologies which are already developed or under development. To start with, we discuss improvements and new technologies in the car and plane system in Chapter 5. This is followed by a discussion of new collective transport systems in Chapter 6.

5 INNOVATIONS IN AUTOMOBILE AND AVIATION SYSTEMS

5.1 INTRODUCTION

Transport modes may be technically characterized by various parameters related to vehicle performance, transport capacity, level of comfort, energy consumption, etc. The automobile system is the only all-purpose transport mode, and this partly explains its success. The user demand for higher quality, government regulations and supply considerations have all led to various improvements of this technical concept which is already more than 100 years old. At the same time the plane is the mode whose use is growing the fastest; this trend is expected to continue in future (see Chapter 3). Both systems seem to have strong behavioural roots in our mobile society. A second reason for choosing car and plane is the fact that these two systems are responsible for most of the global air pollution, and hence for the external costs of transport. Therefore, it is interesting to analyse which improvements and innovations may be expected in these systems, even though the two systems are technically disjoint.

In the development of the private car several trends may be distinguished. The first trend is that existing cars may be improved. Improved automobiles include characteristics such as less weight (except electric cars), improved fuel efficiency and alternative fuel use (including electric and hybrid vehicles). Electric cars especially may be introduced on a larger scale in the next few years. Another development may be in the direction of extended use of electric vehicles and eventually physical guiding of vehicles (the guided vehicle could become a reality). In planes less radical innovations may be expected; besides improvements of current technologies (which ultimately are expected to have a major impact on the environmental burden of aviation), hydrogen may in the long run perhaps replace conventional fuels.

In this chapter we will mainly discuss possibilities to improve the car and aviation system. We particularly concentrate on the former, because the private car is the main mode at the moment where most improvements may be achieved. No attention will be paid to behavioural options like reducing speeds (see e.g. Rienstra and Rietveld, 1996), since this is beyond the scope of this book. Subsequently, we discuss:

- likely and potential **improvements** in current car technology;

- the potential of **alternative fuels** in the car system, which may replace fossil fuels;
- more explicitly **electricity** as alternative source of energy, because this may also be applied in several new collective modes (see Chapter 6);
- **hybrid** and **fuel cell** cars, which may offer an attractive alternative for the current conventional car;
- physical and electronic **guidance** of cars in order to increase the capacity of infrastructure and improve the route choice;
- replacing kerosene by **liquid hydrogen** in the **aviation** system.

Chapter 4 debated critical success factors for new technologies, which may be found in several scientific fields. Many technologies discussed in that chapter are still in their infancy, however, while the success and failure factors are often largely similar. Therefore, in the present chapter we apply these success and failure factors only to the electric car, guided vehicles and hydrogen planes.

5.2 IMPROVEMENTS IN CURRENT CAR TECHNOLOGY

The existing gasoline passenger car has an average energy consumption of about 8 to 9 l/100 km. The resulting 'well to wheel' CO_2 emission is about 20 g/100 km. The CO_2 emission of diesel cars is 20 per cent lower than that of gasoline cars; emission reduction measures (low sulphur diesel, exhaust purification) are likely to decrease this difference, however. With LPG (gas) powered cars the 'well to wheel' efficiency at present equals that of gasoline.

The options for reducing energy consumption by conventional cars can be subdivided into measures improving drive train efficiency, and measures reducing tractive effort requirements. The measures involve the application of various features (Rutten *et al.*, 1990; OECD, 1993b; Okken *et al.*, 1993a). The following drive train improvements may be distinguished (figures in brackets denote possible fuel savings):

- lean burn engines: these engines have an air/fuel ratio of 1.4, and can be equipped with an oxidation catalyst;
- stirling engines: these are 'external combustion engines', with low NO_x emissions due to a low temperature; in principle, this technology may be used in combination with all fuels;
- direct injection diesels (15 to 20 per cent compared to indirect injection diesels, but higher emissions of noise and NO_x);
- engine switch-off when possible (4 to 6 per cent);
- continuously variable transmission (CVT) (10 to 15 per cent);
- regeneration of braking energy, e.g. with a flywheel in combination with CVT (15 to 25 per cent in city traffic);

- ceramic diesel engine, reducing thermal losses through cylinder walls;
- ceramic or fullerene lubricants to reduce engine friction.

The following possibilities for tractive effort reduction may be distinguished:

- weight reduction by using aluminum or plastics in car body and components (6 per cent fuel consumption reduction for 10 per cent weight reduction);
- reduction of air drag, by improving aerodynamic styling (5 to 6 per cent);
- low-rolling resistance tyres (3 per cent).

The above-mentioned options can all be categorized as developments towards more advanced combustion engine vehicles. It should be mentioned, however, that it is not expected that all these technologies will be introduced at the same time, while trade-offs are often necessary between several impacts (e.g. the catalytic converter has a positive impact on reducing local air pollution, but increases CO_2 emissions).

Before the year 2000 the energy efficiency of gasoline cars is therefore expected to increase by 15 to 22 per cent. Diesel-powered cars nowadays consume about 7 l/100 km and may become 15 to 20 per cent more fuel efficient; for LPG cars this may be 15 per cent by using these technologies. With currently available best technology, fuel economy can in principle be improved to a fuel consumption of less than 3 l/100 km. This has been demonstrated in the Renault Vesta and Peugeot ECO 2000. Both cars have an improved aerodynamic styling, a weight of about 470 kg due to the use of light materials, low-rolling resistance tyres, and an engine supplying 20 to 27 kW out of $\pm$ 750 cc. As yet, these cars are not marketed, because they do not meet current customers' expectations of vehicle performance (Rutten *et al.*, 1990; OECD, 1993b). Despite these potentials, Okken *et al.* (1993a) estimate the autonomous efficiency improvement of conventional cars by the year 2030 to be only 25 per cent.

It may be concluded that in current technologies large-scale improvements are possible, which may reduce energy consumption and greenhouse gas emissions to a large extent. The bad news is that car use is expected to grow by at least the same rate as the expected reduction in energy use. Therefore, total energy consumption and CO_2 emissions may rise or stabilize, despite these improvements. As a result the emissions may not be reduced sufficiently (Gwilliam and Geerlings, 1994). To achieve large-scale reductions of emissions, therefore, new technologies should be introduced. New alternative fuels may be a possibility in this respect.

5.3 ALTERNATIVE FUELS

It has been mentioned above that transportation has a wide array of environmental costs, so that fuel is not the only cause of the environmental burden of the transport sector. But fuel consumption is certainly a major source of environmental pollution, so that the search for alternative fuel systems is certainly warranted.

The introduction of alternatives for fossil fuels in transport may serve various objectives. First, it may decrease the dependence on oil and on oil-producing countries. Secondly, it may serve to reduce greenhouse gas emissions. Some alternative motor fuels will be evaluated here with regard to the production system, the energy chain efficiency, the direct and indirect environmental impacts, renewability and the infrastructural requirements. Gasoline and diesel have a direct CO_2 emission factor of 73 g/MJ and 89 g/MJ including indirect emissions in mining and refining. The total CO_2 emission factors of alternative fuels are compared with gasoline in Table 5.1.

Table 5.1. CO_2 emissions from alternative motor fuels

	CO_2 emission
Methanol from coal	150 - 170
Methanol CMP	60 - 110
Methanol from natural gas	90 - 95
CNG	80
Electricity	80
Bio-methanol	10 - 30
Bio-ethanol	20 - 80
Zero-CO_2 electricity	< 5
Hydrogen	< 5

Note: Index gasoline = 100
Source: Okken *et al.* (1991)

This table shows that alternative fuels may contribute largely to reductions of CO_2 emissions. The above-mentioned alternative fuels may be subdivided into synthetic fuels, derived from coal, oil or gas, and biofuels derived from biomass. These will be discussed in the next sections (see also DeLuchi, 1993; McKenzie, 1994; Michaelis, 1995).

5.3.1 Synthetic fuels

There are several synthetic fuels under development or in use; here we will discuss methanol and compressed natural gas (CNG).

Methanol has a direct CO_2 emission factor of 70 g/MJ. When applied in a conventional engine, its efficiency is comparable to that of gasoline. It is a corrosive liquid, so that some modifications of the car fuel system (ignition, mixing, tank and pipes) may be necessary. It can be produced in several ways, which largely influence the resulting CO_2 and other greenhouse gas emissions. Of course, when it is produced by coal-gasification, CO_2 emissions may increase; however, it is possible to apply CO_2 removal technologies. The maximum expected conversion efficiency is 63 per cent in the year 2030. Methanol from steam reforming may reduce CO_2 emissions slightly. Its conversion efficiency is expected to increase from 63 per cent in 1990 to 75 per cent in the year 2030. Methanol produced from combined methanol and power (CMP, e.g. once-through coal gasification) also decreases CO_2 emissions; the conversion efficiency is 70 per cent.

CNG has a direct CO_2 emission factor of 56 g/MJ, and a total emission of 66 g/MJ. The efficiency of a gas motor is slightly lower than that of gasoline; compression causes an energy loss of some 3 per cent. A net CO_2 reduction of 20 per cent compared to gasoline can be achieved, as a result of the low carbon content. CNG is stored on board in a 200 bar tank and in various countries a natural gas distribution network is already present. Filling of CNG can either be done at home with a slow-fill compression system, or at centralized facilities with a fast-fill compression system.

5.3.2 Biofuels

There are three different processes available for the production of liquid transportation fuels from biomass (Okken, 1991):

- mechanical (rape-seed oil);
- thermochemical (methanol);
- biochemical (ethanol from sugar cane or corn).

Biofuels from biomass produced in an environmentally sustainable agricultural process are in principle CO_2 neutral, as the CO_2 produced by burning the fuel has been absorbed by photosynthesis. In practice, however, there may be some indirect CO_2 emissions from diesel consumption by tractors; the same holds for CH_4 emissions from fertilizer production and the energy input in the processing of biomass into biofuels.

On EU surplus agricultural areas (10 to 20 million hectares) some 1000 to 2000 PJ of liquid biofuels can be grown with current technologies. This would be sufficient to fulfil 15 to 30 per cent of the current fuel requirements of the European transport sector (the total agricultural area in the EU is 130 million hectares). The annual energy yield from biomass productions is expected to be 180 to 360 GJ/ha, which is small compared to wind energy (1500 GJ/ha/year) and photovoltaic conversion of solar energy (PV) (5000 GJ/ha/year) potentials.

Rape-seed oil is pressed from rape seed, together with by-products of animal feed and straw, which can be pressed into fuel briquettes. It can be used directly or can be esterified with methanol into biodiesel.

The production of bio-methanol from wood or straw is only at an experimental stage. The first step in the production process is thermal gasification of dry biomass. From this synthesis methanol is produced by reforming or catalytic synthesis. The overall CO_2 emission factor of bio-methanol may be better than that of gasoline. The conversion efficiency is estimated to be some 55 per cent (Okken *et al.*, 1993a). A problem may be that current investment costs using this method are 3.5 times higher than that of methanol production from natural gas.

The production of bio-ethanol from corn or sugar cane is a proven technology. It is produced from starch or sugar ethanol in a biochemical fermentation process, which results in a mixture of water and ethanol, from which ethanol is obtained by distillation. Both fermentation and distillation have high energy demands, but these can be improved considerably. Ethanol can be mixed with gasoline (up to 10 per cent) without requiring any adaptation of the engine.

A huge advantage of (bio)methanol and (bio)ethanol is that the distribution and filling infrastructure closely resembles that of gasoline and diesel; so it may be set up parallel to that system.

5.3.3 Hydrogen

Hydrogen (LH_2) can be produced in different ways (Okken, 1992). It can be produced from coal and steam by means of a shift reaction between CO and water, in combination with the production of CO_2 and substitute natural gas or medium caloric gas. The overall efficiency of this process is 70 to 75 per cent, of which 55 per cent is H_2; CO_2 removal is possible. Hydrogen can also be obtained through steam reforming of natural gas or oil, with an efficiency of 75 per cent at present, probably increasing to 80 per cent in 2030.

Hydrogen can also be produced by electrolysis of water. This process produces only oxygen and hydrogen, and is free of indirect CO_2 emissions if the consumed electricity is produced from nuclear energy, hydroelectric power or PV. In case of production by PV in the Sahara an efficiency of even 84 per cent may be achieved. Investment costs are 50 times higher than for steam reforming with natural gas, but low maintenance and operational costs as well as a long lifetime may make this system competitive with other technologies, when the other systems are equipped with CO_2 removal devices. According to Okken (1992), production of H_2 from steam reforming of CH_4 with CO_2 removal is cheaper than from electrolysis with PV electricity. The current production costs are about $20 per GJ of hydrogen produced (Okken, 1992).

Hydrogen can be stored under pressure in the gas phase, or as a liquid in a cryogenic storage tank at a temperature of 20 K and a pressure of 2 bars. Hydrogen stored in a cryogenic container evaporates at a rate of 0.5 to 2 per cent per day depending on the quality of the thermal isolation. As hydrogen/air mixtures can be highly explosive, special safety precautions (ventilation) are necessary. Hydrogen can also be absorbed in metal alloys, forming a metal-hydride. However, the high weight and low storage capacity of these systems are, at least for the time being, prohibitive for practical application.

For liquefaction a significant amount of electricity is required. Present-day systems consume about 0.4 MJ per MJ hydrogen, which may be improved to 0.29 MJ/hydrogen, by applying integration of heat flows and cascading. In the liquefaction process a large flow of low-temperature heat is produced, which may be used for space-heating. In this way 50 per cent of the energy losses may be recycled (Okken, 1992).

In transport applications hydrogen can be used in adapted car engines, in gas turbines or in fuel cells. On-board storage in cryogenic tanks offers the highest energy density per unit volume, so that this storage system will most likely be used for cars or planes. Storage in pressure cylinders may be used in larger road vehicles such as trucks and buses. In distribution and filling of hydrogen, evaporation losses of 10 to 30 per cent may occur. If the evaporated hydrogen can be recycled, only the liquefaction energy is lost. The possible application of hydrogen in planes will be discussed in Section 5.9.

The distribution of gaseous or liquid hydrogen requires a completely new infrastructure, which may be incompatible with the existing organization of fuel distribution. Constructing a new pipeline network, in combination with compression and/or liquefaction facilities, will be expensive. Introduction of hydrogen will therefore probably occur firstly in applications where large filling stations can be combined with a compression/liquefaction unit (e.g. at airports for hydrogen planes).

5.4 TRENDS AND FUTURE DEVELOPMENTS IN ELECTRICITY GENERATION

The current car and plane systems are fuelled largely by fossil fuels such as petrol, diesel and kerosene. Trains, trams and metros are operating on electricity, which is produced by using oil, gas, coal, nuclear energy and sometimes renewable sources. To consider future possibilities for reducing greenhouse gas emissions, it may also be attractive to use electricity to some extent in the car system. To evaluate these possibilities, it is first necessary to discover what future electricity generation will look like. First, possible and expected improvements in electricity generation are discussed, which may also be important for the future current and collective modes which are powered by electricity (see Chapter 6). Then we turn to the prospects for electric, hybrid and fuel cell cars, respectively.

5.4.1 Future generation of electricity

The primary energy consumption and total emissions of electric transport systems depend on the composition of the electricity generation system. This composition varies from country to country. In this section we will restrict ourselves to an evaluation of trends and future possibilities for the Dutch situation, though the scenarios discussed may also be applicable to other countries.

For reducing CO_2 emission from electricity generation basically four different routes are available:

- CO_2 removal from fossil fuel plants, e.g. from integrated coal gasification combined cycle (IGCC) plants and storage in, for example, depleted natural gas reservoirs;
- electricity production by using renewable resources in thermal plants, based on closing the CO_2 cycle (biomass);
- electricity production from non-carbon-based renewable resources. These options comprise solar energy (using PV or photothermal conversion), wind power, hydroelectric power, wave and tidal energy, geothermal energy, etc.;
- nuclear generation.

Most technologies for using renewable resources are still at an early stage of development, so that these technologies are not yet economically attractive. Some technologies may become viable in the next two decades (wind energy and some PV-applications), but unless fossil energy prices increase significantly most technologies will not or will only marginally become competitive before 2040 (Okken *et al.*, 1991 and 1993a). The present and projected composition of the Dutch public electricity generation system is presented in Table 5.2.

The acronyms ER, BG and BGvar in this table represent three different scenarios, based on two different socio-economic forecasts (ER = European renaissance and BG = balanced growth; see Section 3.7.1), where the BGvar scenario places extra emphasis on CO_2 emission reduction compared to the BG scenario. Projections based on a macroeconomic optimization under various CO_2 reduction constraints are *inter alia* presented in Okken *et al.* (1991, 1993a and 1993b).

Table 5.2. Present and projected future composition of public electricity generation in the Netherlands (per cent)

	1990	2000 ER	2015 ER	2015 BG	2015 BGvar
coal	38	30	10	18	29
gas	42	44	62	47	28
nuclear	5	5	0	14	30
renewable	1	8	15	14	11
import	14	13	13	7	2

Source: Boonekamp *et al.* (1992)

It may be concluded that in 2015 up to 15 per cent of the electricity may be produced from renewable resources. Some 50 per cent of the installed capacity of renewable sources may be wind turbines. Incineration of waste and biomass will make up 25 per cent of the installed capacity and solar energy (PV) will make up the remaining 25 per cent (mostly roof-mounted solar panels, which may become economically attractive around 2015). It should be noted that in other countries the composition of electricity production may differ a great deal. France, Belgium and Germany, for instance, have much more nuclear generation of electricity, while countries like Norway have a much larger share of renewable sources.

5.4.2 Future CO_2 emissions by electricity generation

The expected and future efficiency and emissions of electricity generation in the Netherlands are presented in Table 5.3.

It appears that a significant increase in generation efficiency is clearly possible in the medium term. Without nuclear energy, a reduction of the CO_2 emission per kWh_e of 14 per cent is foreseen in the ER-scenario. This reduction mainly results from an increasing use of natural gas. In the BG- and BGvar-scenario the share of nuclear energy is increased and, due to an assumed CO_2 tax, the application of CO_2 removal from coal gasification plants in combination with storage in depleted natural gas reservoirs becomes economically attractive. In these scenarios a CO_2 emission reduction of 42 to 60 per cent is expected.

Table 5.3. Efficiency and emissions of electricity generation in the Netherlands (g/kWh$_e$)

	Eff. (%)	CO_2	NO_x	SO_2	CO	HC	part.
1992	39	646	1.09	0.48	0.025	0.005	0.031
2000	39	690	0.61	0.31	0.025	0.005	0.016
2015ER	42.5	596	0.43	0.12			0.008
2015BG	48.5	398	0.35	0.06			0.006
2015BGvar	56.7	269	0.29	0.06			0.006

Source: Boonekamp *et al.* (1992)

Calculations in Okken *et al.* (1991) show that CO_2 emissions from central electricity production in 2020 may decrease to ± 150 g/kWh$_e$ in a scenario without nuclear energy. This decrease of almost 80 per cent is realized by the use of wind energy, PV and fuel cells. In a scenario with nuclear energy and with the same reduction constraint, the CO_2 emissions from electricity production may become as low as 7 to 11 g/kWh$_e$.

In a base-case scenario without CO_2 constraint, the only renewable resources that become economically feasible are biogas, land-based wind power, hydroelectric power, geothermal heat and ethanol (in the transport sector).

5.5 ELECTRIC CARS

The technology to improve cars that is nearest to introduction is the battery-electric vehicle (electric car). Several models are sold on the market already, although these do not yet seem to be a real competitor for conventional cars. The development of the technology goes fast, however. Several states in the United States are specially promoting the sales of these cars. California, for example, has a regulation that requires 2 per cent of the cars sold to be electric in 1996; in 2000 this percentage should be 10 per cent.

Therefore, it is of interest to analyse this technology in more depth. First, we will discuss the technical features of this mode. Next, we will apply the success and failure factors (as discussed in Chapter 4) on this technology.

5.5.1 Technical factors

The drive train of an electric car basically consists of a battery for energy storage, an electronic engine controller and an electric engine. An on-board or out-board charger serves to charge the battery with energy from the electricity grid. During braking the electric engine can act as generator to recuperate braking energy, which is stored in the battery (Smokers and Van Hilten, 1992; Smokers, 1993; Van Hilten *et al.*, 1992). Battery-electric vehicles are emission-free during driving. The emissions associated with electricity generation have been discussed above.

Electric cars may be introduced in a number of market niches. Due to the low energy density of presently available batteries the range of such vehicles is limited to 70 to 100 km, while the current top speed is about 100 km/h. Experimental vehicles, however, have proven top speeds of 200 km/h and ranges over 200 km. Pb-acid batteries are used in most vehicles because of their low costs, high power density and mature technology. NiCd-batteries are expensive, but have nevertheless been applied in several vehicle types. Due to their limited range the use of electric cars will for the time being be restricted to urban traffic.

The final energy consumption of current electric cars is ± 300 Wh/km. Light-weight vehicles (mostly two-seaters) consume ± 200 Wh/km. Developments between 2000 and 2015 may result in full-size but light-weight electric cars with a final energy consumption of 100 to 150 Wh/km.

Intensive research is being undertaken to develop an advanced battery type that may increase the range of electric cars to values comparable with conventional cars. In Table 5.4 the projected characteristics of some promising options are presented. A problem with most advanced options is that their energy density may exceed that of the conventional Pb-acid battery, but their power density usually does not. The minimum power required to propel a medium-weight passenger car is of the order of 25 kW. The maximum power density in Table 5.4 is 200 W/kg, so that in order to meet this power demand at least 125 kg of batteries are required.

Since the energy consumption of the passenger car considered is about 100 to 150 Wh/km, even with the most optimistic battery characteristics of Table 5.4 the range is still limited to 170 to 250 km. Increasing the range can be realized by using a larger battery, but the battery with the characteristics assumed above will not be cheap. The current generation of electric cars is generally equipped with 300 to 500 kg of batteries making it significantly heavier than comparable conventional cars. Another problem may be that the battery life is limited. This means that the battery will have to be replaced up to three times during the lifetime of the vehicle.

The battery of electric cars has to be charged from the grid. Charging a 20 kWh battery from a 3 kW household outlet will take at least 7 hours. There are also technologies which charge a car in about 20 minutes, but these are very expensive and shorten the life of the battery. Furthermore, this is not possible for all kinds of batteries (Martens and Boonekamp, 1990). If charging is to be comparable to filling the gasoline tank of a conventional car, charging times of less than 15 minutes are required. This would require 80 kW charging stations.

Table 5.4. Projected characteristics (2010/2015) of some battery types

	Energy density (Wh/kg)	Power density (W/kg)	Eff. (%)	Life-cycle	Costs (fl/kWh)
Pb-acid	50 - 60	150 - 200	85	1000	200
NiCd	60 - 70	200	80	2000	500 - 750
NiMH	50 - 70	175	80	1000	500 - 750
NaS	140 - 170	150 - 180	91	1000	500 - 800
NaNiCl	110 - 140	100 - 150	85	1000	200 - 400
Li-polymer	150 - 200	150	95	800	300 - 600
ZnBr	70 - 75	70 - 90	75	1000	250

Source: Okken *et al.* (1991 and 1993a)

Application of such systems requires adaptations of both the electric circuitry in the car and of the grid, which will also make safety measures necessary.

Small-scale introduction of electric cars will require little infrastructural investment, as the electricity grid is already present. For large-scale introduction public charging stations will be necessary, with resulting investments in grids and facilities.

5.5.2 Spatial factors

The most important negative critical spatial feature of electric cars is the small maximum distance which they can bridge. Therefore, electric cars are likely to be used in cities only. In the 'compact' city, land use and transport planning largely favour the use of collective modes, and there might be little space available for the electric car, which may hamper its introduction. When, however, additional attention is given to road infrastructure and parking facilities at employment sites, the option of improved (short distance) cars may well be feasible. This is certainly the case in a diffuse city, where traffic flows are relatively thin, which makes the use of individual modes attractive.

5.5.3 Institutional factors

The present private car system is firmly embedded in society. There is a market for innovations, however, since there is a certain degree of control by the government and a growing awareness of the social importance of innovation. Governmental support may be high therefore (as is the case in California). The necessary adjustments of the road infrastructure are limited, which may be an important success factor. Since it is an individual mode, there may not be many other failure factors in the institutional field.

5.5.4 Economic factors

Analysis of the market for electric cars

At the moment several electric cars are on the market; mostly these are experimental cars and their costs on the market are high. The most important advantages for the users of these cars are the low energy costs, which are only 1 to 2 cents per kilometre. Fuel costs of conventional cars are about five or six times as high. In this case, it is important to note that there are high taxes on fuel, but not (yet) on electricity. Another advantage is the lower maintenance costs, because the engine is simpler and cleaner. From this it follows that the variable user costs are very low compared to those of conventional cars.

In particular the range and the fuel costs may have a significant impact on the market penetration (Bunch *et al.*, 1993). One of the main problems therefore is the low range, while the price is about twice as high as that of conventional cars. It is expected that this price will be lowered after some years because of returns to scale, when production is growing. The price of an electric car without battery will probably become about as high as that of a conventional car. The problem, however, is the price of the batteries: the cost of a battery with a storage capacity of 20 kWh, for example, can be calculated from Table 5.4, and may range from \$2500 to \$10000. The life of these batteries is also much shorter than that of the car, so they must be renewed a few times. The cost of the batteries could become somewhat lower in the future depending on the development of the technology, but their costs will probably remain high, because the energy capacity of the batteries will have to increase in order to increase the range and speed.

Another problem is that infrastructure needs to be constructed for charging the cars. Especially in urban areas there is in many cases not enough space to load the car at home, especially in high-rise areas. There might also be problems with the capacity of the electricity network. What is clear, is that for this infrastructure much capital is needed (Martens and Boonekamp, 1990).

The market for electric cars consists of people who use their car (almost) only for short-distance trips, for example, to go to their job every day. The problem is that almost everyone uses their car from time to time for longer distances also (holidays, visiting family, etc.). If people only have an electric car, long-distance trips are not possible. Therefore, the electric car might be especially attractive for households who also own a conventional car.

The biggest (sub)market will be found in the larger urban areas, where people do not drive longer distances most of the time. Schol and Smokers (1993), for example, calculated a market potential for electric cars in Amsterdam for the year 2015 of 68 per cent of the private cars, 72 per cent of the delivery vans, 40 per cent of the trucks and 80 per cent of the buses. In smaller cities and rural areas this potential will be much lower, however.

Financing electric cars

An important condition for the introduction of electric cars is that the costs should not be much higher than those of competing conventional cars. A government incentive scheme is needed, therefore. The government may well want to do this, because these cars are at the local (urban) level emission free, while at the macro-level emissions are lower (depending also on the way electricity is produced). Another important positive external effect is the reduction of noise nuisance.

The variable costs (i.e. without battery costs) of electric cars are a problem, because they are very low. Therefore the mobility of the owners will probably increase, which may cause congestion problems, especially in already choked urban areas. A solution for this problem would be for the government to increase the variable costs and reduce the purchase costs.

Another problem is the financing of the charging infrastructure. In principle the charging infrastructure might be financed by the private sector, like conventional petrol stations. The actual building of this infrastructure will probably have to be government backed. Otherwise, the construction may proceed slowly in the beginning, which would make electric cars unattractive during the start of large-scale introduction.

5.5.5 Socio-psychological factors

Theoretically, the electric car may be seen as an improvement of existing cars. The acceptance of electric cars may be high in this case, because of the positive environmental impacts and because no extensive new infrastructure is needed. One problem might be the lack of space in cities, which may cause resistance to large numbers of cars in urban areas.

The adoption may be low in practice because of the high costs, the low range of the electric car and its low speed, which makes this car unattractive for conventional car users.

5.6 HYBRID CARS

Hybrid-electric vehicles (hybrid cars) may be able to combine some of the benefits of electric traction with the range, performance and fast fuelling characteristics of conventional cars (see Okken *et al.*, 1991; Okken *et al.*, 1993a). The drive train of hybrid cars contains an electric part with an energy storage device (in most cases a battery), a controller and an electric engine, as well as a thermal engine. There are two basic types of hybrid car: parallel configuration and series configuration.

In the parallel configuration either the electric or the thermal power sources drive the wheels via a transmission which is able to switch between both power sources, or which may even be able to combine power from both sources. In general, the electric power source is used for emission-free urban driving, while the thermal engine is used in situations where high speed or high power is required (e.g. highways). The thermal engine can be a conventional car or diesel engine, but may also be a stirling, two-stroke or lean-burn engine, and can operate on conventional or alternative fuels. Battery volume and electric motor power will be small, so that speed and range in urban traffic will be less than that of an electric car. The overall energy consumption will be comparable to that of a conventional car.

In the series configuration the thermal engine acts as a generator providing electrical energy to the battery, which in turn feeds the electric motor. If the thermal engine is small, it merely acts as a range extender providing some battery charge-up during driving. A larger engine can serve as the main power source, making operation independent of the grid possible. The range is comparable to that of a conventional car. The thermal engine can be a conventional or an advanced car, but also, for example, a gas turbine, and may operate on conventional or alternative fuels. Battery weight can be limited if a high-power battery is used. The electric energy storage system in hybrid cars may also be a flywheel or supercapacitor.

The advantage of the series configuration is that the thermal engine can be smaller than in a normal car, because no peak power production is needed. Instead it is operated at constant, nominal power, where its fuel efficiency is optimal and emissions are low. For experimental conversion cars fuel consumption of 4 to 5 l/100 km has been demonstrated. Without further modification of car design, fuel consumption of series configuration hybrid cars may thus be 25 to 50 per cent lower than that of a comparable conventional car. When hybrid cars are made from light-weight materials, energy consumption figures may be as low as 1 to 1.5 l/100 km (Lovins and Barnett, 1993). Therefore, from the point of view of energy efficiency and reduction of greenhouse gas emissions, hybrid cars seem to be very promising.

Hybrid cars are currently under development by most of the large automobile manufacturers, while some models are already produced in small quantities. They may become commercially attractive within the next decade. Due to the double drive train, hybrid cars will be more expensive than comparable conventional cars.

If hybrid cars are fuelled with conventional fuels, they can make use of the existing fuel distribution system, and no infrastructure construction will be required. Parallel hybrid cars will require grid-connected charging facilities.

5.7 FUEL CELL CARS

Fuel cell powered vehicles (fuel cell cars) are a kind of hybrid car in the sense that they also have an electric drive train combined with an on-board power source (see Okken *et al.*, 1991; Okken *et al.*, 1993a). In this case, however, the power source is not a thermal engine but a fuel cell.

A fuel cell is an electrochemical device which directly converts chemical energy from fuel into electrical energy. There are various types of fuel cell available, but for transport applications the solid polymer fuel cell (SPFC) is the most appropriate. At the anode a SPFC is fed with hydrogen which is ionized at the surface of a catalyst. Hydrogen ions (H^+ or protons) can travel through a polymer membrane to the cathode where they react with oxygen at the surface of a catalyst to form water.

An SPFC thus produces electricity from the catalytic oxidation of hydrogen, emitting only water. The conversion efficiency may be 55 to 65 per cent. An advantage of fuel cells is that their efficiency is higher for partial loads. Hydrogen can be supplied to the SPFC from an on-board hydrogen storage, or from an on-board reformer which converts methanol or CNG into H_2 and CO_2 (DeLuchi and Ogden, 1993). On-board storage as well as the filling of methanol may, however, be easier than that of hydrogen, while methanol may also be cheaper.

Current prototypes and concept designs show an SPFC, which is operated at constant power, so that a battery or flywheel is required for storing electrical energy and providing peak power. In the future the response time of SPFC systems may become shorter, and therefore the SPFC may be operated in a transient mode, directly supplying energy to the electric engine.

The overall system efficiency of fuel cell powered vehicles can be 40 to 50 per cent, so that the final energy consumption may be 30 to 50 per cent less than for a conventional car. In calculating the primary energy consumption the efficiency for hydrogen or methanol production, storage and distribution have to be taken into account.

At present the price of an SPFC is about US$1550/kW. With future mass production this price is expected to decrease to US$75-150/kW. Currently field tests with SPFCs in a bus are being carried out and other bench and field tests are planned. Applications in large vehicles may become viable within 10 years; passenger cars may become available in 10 to 20 years.

5.8 GUIDED CARS AND TELEMATICS

Even if we narrow down the meaning of 'guided vehicles' to vehicles for individual transport (in contrast to collective modes), it is a rather broad description which may comprise two entirely different systems: *physically* and *electronically* guided vehicles.

Though the development of physically guided vehicles started as far back as the 1960s, it is still not clear yet when such a system will actually be introduced. It is also not yet clear what the system will look like. One of the most developed systems is called Taxi 2000. This is an urban transportation system, designed for specific urban areas (Van den Berg, 1993). It is a system of minimum size and minimum weight vehicles that operate under automatic control between stations in a network of narrow guideways. In each passenger vehicle three adults may be seated. Empty vehicles can be ordered continually and automatically, so that they can anticipate demand and wait for people rather than having people wait for vehicles. Both passenger and freight vehicles may operate on the same network.

The application of telematics to road transport may have huge consequences in the next few decades. Electronic guidance may be anything between route information systems, external speed control, train formation on highways, and fully automated route guidance. The most simple form is information given by radio stations, the most complex form is a system with continuous interaction between driver and information system. In this section we limit ourselves to some remarks about telematics in general, since these technologies may be applied within all the modes discussed above, although the impact on greenhouse gas emissions may be limited. For an extensive discussion on technical, social, economic and market aspects of telematics in the transport sector, the reader is referred to Nijkamp *et al.* (1996b). The general conclusion from the latter study is that telematics offers interesting opportunities, but that its problem-solving capacity regarding environmental issues is often overrated.

5.8.1 Technical factors

Physically guided vehicles

With physically guided vehicles the interaction with the guideway may be mechanical (like the rails of a train), or electromagnetic (see also the discussion on Maglev systems in Chapter 6). If the vehicles and their centrally operated guideway infrastructure are inseparable, i.e. if the vehicles are bound to the guideway and cannot be operated autonomously, such a system may be a kind of blend between a collective and an individual transport system. Passengers in these systems can travel individually in small vehicles, and can go to any destination at any given time, and so be independent of time schedules usually found in collective transport.

In 'roadway powering' systems the guided vehicles are also capable of autonomous driving separate from the guideway. On normal roads the vehicles operate like regular passenger cars propelled by an on-board energy source. On powered road sections (e.g. highways) the vehicles receive their energy from the road and their on-board energy source (conventional, battery, or whatever) is switched off. Energy transfer may be by linear induction (see Section 6.3), but other systems have also been proposed.

Physical train formation of autonomous road vehicles does not seem very attractive from a technical point of view. When establishing a physical connection between (driving) vehicles, there is always a speed difference which will cause a sudden acceleration or deceleration at the moment of coupling. This shock will have to be absorbed by the cars' bodies. This would require a stiff body structure or chassis in the longitudinal direction (direction of motion). Such a structural demand is incompatible with safety features like deformation zones and shock-activated airbags.

Electronically guided autonomous road vehicles
Developments towards electronic guidance of cars have, for example, been made in the US, Europe and Japan. A well-known US program is Intelligent Vehicle Highway Systems (IVHS) (OECD, 1992), in the EU there is the DRIVE-program (see e.g. Nijkamp and Pepping, 1995). Many systems can be imagined, which vary from very simple ones (electric road signs, radio broadcasting) towards systems in which communication is required between the vehicle and a computer and communication system, through which information on vehicle position and speed is exchanged and control command can be given to the vehicle (Jeffery, 1990). On highways the guidance system can equalize speeds of and control the distance between vehicles, so that a form of contactless train formation is achieved. If all vehicles are centrally controlled, the distance between vehicles may decrease and speed may increase. This would benefit traffic flow considerably and increase highway capacity.

Electronic guidance systems in general may increase fuel efficiency due to the enforcement of rationalized driving behaviour and efficient route selection. Therefore, they may contribute to the reduction of CO_2 emissions by transport (Shladover, 1993).

5.8.2 Spatial factors

An important, positive feature of a guided vehicle system is the relatively small use of land. Furthermore, physically guided systems with inseparable vehicles are designed for relatively small distances where sufficiently dense passenger flows are to be transported. Good examples are airports and large business centres. Accordingly, when cities develop in a more 'compact' way, the systems may be feasible on particular high-density trajectories. The use of systems like Taxi

2000 seems to be unrealistic in diffuse cities and on higher spatial scale levels than cities, where distances are longer and passenger flows more spread.

Telematics systems claim a small amount of extra land for infrastructure. From this point of view, there seems to be no restriction to their introduction. The systems themselves may, however, influence the spatial organization (Soekkha *et al.*, 1990). When we come to the future pattern of urbanization on various scales, all patterns which generate traffic in relatively dense bundles are likely to be introduced rapidly. On the metropolitan level, this means that 'compact' cities have a higher chance of (a fast) introduction than diffuse cities where a more spreadout traffic pattern dominates. On the interurban (national) and international level, the urbanization model may not be very important because heavy congestion may occur in central parts of both chains and zones patterns and specialization and concentration patterns.

5.8.3 Institutional factors

A main problem is the lack of standardization in information systems, which prevents the benefits of telematics from being exploited to the full. In this respect, governments might have to play a more active role. Pressure from national industries might also be influential in the ultimate choice of specific systems and standards.

5.8.4 Economic factors

Analysis of the market for information systems
Because little is known about the economic factors involved in physical guidance systems or the costs of the many electronic ones that are available, in this section we concentrate on the potential benefits of such telematics systems (further details can be found in Emmerink, 1996).

The benefits of information systems are to be found in the more efficient way of travelling for both individual and collective transport modes. Therefore, travelling times and average distances will be shorter and the average speed higher. A more efficient way of travelling is achieved when drivers are able to improve trip-related decisions. This holds not only for departure time and destination choice, but also for decisions related to trip scheduling, parking choice and destination choice. Even higher level impacts, like shifts in mode choice and change in behaviour should be taken into account. By using an information system, time and also energy costs will be saved, and the infrastructure may be used more efficiently and the number of accidents may decrease. The precise revenues from information systems are not easy to assess. Several studies have been carried out on this subject, which estimate the savings in travelling time to be from 4 to 20 per cent, depending on assumptions about the market penetration and the system itself (Emmerink *et al.*, 1994; OECD, 1987a and 1988a).

Potential users of such systems will be the drivers who are attracted by the savings in time and fuel. Business travellers may make most use of telematics, because they have a high time value, drive a lot and frequently use unknown routes. Recreational drivers may not be willing to pay high prices for such systems, because they drive less and have a lower value of time.

Financing information systems

As mentioned above, many information systems are possible, depending on the use of different types of telematics. It is clear that these systems may lead to huge savings not only for users, but also for society. A large part of these benefits is found in a reduction of negative external effects (less air pollution and congestion), but the government also benefits (less maintenance, accidents and new infrastructure). And users, too, can benefit directly from these systems (lower fuel costs and time savings).

So users may be prepared to pay for the information systems as much as they benefit from them. Studies report that car users are interested: for example, more than half of them are interested in a dynamic real-time system. It is clear, however, that the market penetration will depend to a large extent on the price (Emmerink *et al.*, 1994).

The costs vary considerably, depending also on the characteristics of the system, especially of the technology used. At the moment it is unknown how returns to scale will influence the price. It is clear, however, that large-scale introduction will reduce the price per user.

Most studies also suggest that the costs (including the infrastructure needed) will be higher than the direct revenues for most users. So the government may see it as its task to support the development of telematics technology and the introduction of information systems. A problem may be the rise of mobility, because driving becomes more attractive when it is possible to avoid congested sections. So it is not at all certain whether the government will actually support the introduction of telematics, as is evidenced by various recent examples in Europe. Another task of the government may be to choose one or just a few systems, as it is not efficient to have too many systems if returns to scale are to be achieved.

A solution may be that the government or an organization which is tied to the government finances the system. A user charge may then be levied, which is dependent on the level of demand. Probably the (socially) optimal charge will not cover all costs, but part of the investment might be paid back.

5.8.5 Socio-psychological factors

Little can be said about physical guidance systems in terms of acceptance and adoption. Information systems will probably have little acceptance problems, because no new infrastructure is needed and there are environmental advantages, as well as advantages for the users.

The adoption will mainly be dependent on the costs of the systems, which will have to be weighed against the advantages. A problem may be the complexity of the systems, which may discourage potential users.

Having discussed various future options for new car technologies, we move on to a short discussion of a new technological breakthrough in the aviation sector.

5.9 HYDROGEN PLANES

A new technology in transport is the development of hydrogen engines (see Section 5.3.3). Such hydrogen engines may be used in both cars and planes; the types of engines used are very different, however (see also DeLuchi and Ogden, 1993). The technology is still in the R & D stage and hence it is not yet clear whether hydrogen can be introduced on a large scale. The biggest advantage of hydrogen is the low environmental impact, because it generates almost no air pollution. Therefore, it might in principle be interesting to replace conventional engines with hydrogen engines.

In aviation - responsible for about 3 per cent of the global CO_2 emissions - hydrogen is one of the very few options for reducing CO_2 emissions. Deutsche Aerospace Airbus Gmbh (DASA) and some Russian firms are developing a prototype hydrogen passenger plane, the cryoplane (Kemp, 1994). A few experiments have been carried out with liquid hydrogen planes, e.g. with an experimental Tupolev Tu-155 in 1988 in Russia (a three-engine plane with one engine operated on hydrogen) (*Polytechnisch Weekblad* 7/8, 1991; *Intermediair*, 1993). After the year 2005 hydrogen planes might be introduced more extensively. The next sections concentrate on the application of hydrogen in planes (see also Hoogenstrijd, 1993).

5.9.1 Technical factors

Technical features

The density (kg/m^3) of liquid hydrogen is extremely low (8.5 per cent of the density of kerosene), so that very large cryogenic storage tanks are needed, which requires a totally new plane design. Storage tanks in the plane body (e.g. above the passenger compartment) seem to be the most feasible solution. It may be assumed that the hydrogen plane may cost the same as a conventional plane in the future (Okken, 1992).

The lifetime of a plane is roughly 25 years, so that penetration will be slow. At most one sixth of the total fleet can be replaced every 5 years. If introduced in 2005, full replacement is possible by 2030 (Victor, 1990).

Introduction of hydrogen planes will require the construction of a hydrogen production, storage and distribution infrastructure, which for some time will have to exist parallel to the kerosene infrastructure.

An important advantage of hydrogen is its high energy density (J/kg), which is a factor 2.7 higher than that of kerosene. Although hydrogen storage tanks are heavier than kerosene tanks, the start-weight of the plane can be decreased by 33 per cent with respect to the conventional plane (Spee and Wittenberg, 1978), reducing final energy consumption by some 12 per cent (Okken, 1992). Together with an efficiency of 70 to 80 per cent for conventional hydrogen production, the primary energy consumption of a hydrogen plane roughly equals that of a conventional plane.

In Victor (1990) very optimistic efficiency forecasts are presented. For hydrogen planes a possible efficiency improvement of 50 per cent for new conventional planes is expected, while an additional reduction of energy consumption by about 33 per cent may be achieved between 2000 and 2050.

Greenhouse gas emissions

In Table 5.5 the greenhouse gas emissions are presented for fuels which may be used by planes. The emissions of NO_x can be reduced by more than 70 per cent by the introduction of a reduced emission combustion chamber (Grieb and Simon, 1990).

Table 5.5. Emissions per 10 MJ of consumed fuel for planes

	Kerosene	Hydrogen	Methane
Water (kg/10 MJ)	0.29	0.74	0.44
CO_2 (kg/10 MJ)	0.70	0	0.55
NO_x (g/10 MJ)	4.1	6.1	3.4
CO (g/10 MJ)	0.4	0	0.2
Hydrocarbons (g/10 MJ)	< 0.2	0	< 0.2

Note: 10 MJ is equivalent to 0.3 l of kerosene and 1.2 l of hydrogen.
Source: Grieb and Simon (1990)

In the form of clouds, water is a very efficient greenhouse gas. The use of hydrogen planes at high altitudes may lead to the formation of long-lived condensation trails, which may possibly exacerbate global warming. It is estimated that the contribution of condensation trails to global warming may be as high as 5 per cent. Emissions of water in the stratosphere may also lead to further depletion of the ozone layer. These objections may be alleviated somewhat by flying below the stratosphere (i.e. lower than 10 000 m), as at these altitudes water vapour diffuses much more quickly.

Another important external problem of air transport is noise nuisance, which will not be solved by the introduction of hydrogen planes. Furthermore, the application of hydrogen planes requires special safety measures as mixtures of hydrogen and oxygen gas are highly explosive, and therefore leakages must be avoided.

5.9.2 Spatial factors

A negative feature of hydrogen planes is that their introduction will require the construction of a completely new hydrogen production, storage and distribution infrastructure, which is incompatible with the existing infrastructure of kerosene. Both kerosene and hydrogen fuel systems will have to be in operation simultaneously for a certain 'transition' period. This requirement may put a too heavy pressure on land in and around airports. At the same time strong safety measures for distribution and storage seem to be necessary on a permanent basis, which may require additional use of land.

5.9.3 Institutional factors

The infrastructure and organization of the existing aviation system would probably be largely maintained, if the hydrogen plane is introduced. However, the technology requires new general and technical services with regard to the energy supply. The government's role may be limited, since aircraft are largely operated by the private sector. As there is no question of construction of new line infrastructure in air transport, hardly any or no coordination problems are to be expected. Possible institutional barriers are to be found in environmental legislation. Hydrogen requires large-scale storage, transfer and transport of the fuel. The question is to what extent safety problems may arise and to what extent legislation is introduced to cope with these problems.

5.9.4 Economic factors

The biggest problem of hydrogen is the high costs, compared with conventional fuels. The energy costs for a hydrogen car are calculated to be 100 to 500 US cents per kilometre, whereas conventional fuel costs are 5 to 10 cents. The technology may become cheaper, probably when it is introduced more widely, because of returns to scale.

There are also problems with the costs of the infrastructure needed for production, storage and distribution; new extensive infrastructure is needed, which will need huge investments. This will also cause many safety and environmental problems that will have to be solved before the introduction of this type of engine.

Because of the low external effects of this engine, governments will probably support the research and introduction of such an engine. It is not clear, however, whether this technology will develop as a real competitor for conventional engines.

5.9.5 Socio-psychological factors

The introduction of a liquid hydrogen plane may be seen as an improvement on the existing plane. The acceptance in society will probably depend on the safety aspects of the fuel (recall the historic Zeppelin disasters). The same reservation holds, together with the costs, for the adoption of the system.

5.10 CONCLUSIONS

The two most favourite transport modes, the car and the airplane, cause the highest environmental stress. There appear to be many possibilities for reducing externalities caused by the car and aviation system, which may be subdivided into improving the current technology and introducing new alternative fuels. As far as the first option is concerned it is expected that these developments will not be sufficient to reduce, for example, the CO_2 emissions meaningfully, while the introduction of new fuels is not yet practicable on a large scale. Neither of these options offer any solution for congestion problems. In this respect, it may be interesting to analyse which collective modes are now under development and may be introduced in the next decades. This will be the subject matter of the next chapter.

6 THE POTENTIAL OF NEW COLLECTIVE TRANSPORT MODES

6.1 INTRODUCTION

In the previous chapter improvements and innovations of the car and aviation systems have been discussed. Another possibility to reduce the externalities caused by transport, however, is a modal shift towards collective modes (see also Dimitriou, 1994). These modes - though not harmless either - cause less externalities than the current car and plane systems (see Chapter 2). We may subdivide the most important collective options into train, metro, tram and bus systems. However, the future emissions of these modes mainly depend on fuel efficiency and thus indirectly on the future emissions of electricity production, which has already been discussed in Chapter 5; for the bus also alternative fuels may be introduced. Therefore we will not discuss any of these existing modes in the present chapter, but instead concentrate on several new collective modes that have been developed, which will or might be introduced in the coming decades.

There are three main options: the high-speed train, Maglev low- and high-speed applications and subterranean systems. The selection is based on current technological concepts; however, it may well be that these concepts change in the future. The same success and failure factors of the modes, which were discussed in Chapter 4, will be applied here as well. The notion of a 'transport system' covers not only the transport infrastructure and the vehicles, but also the management of the system.

6.2 THE HIGH-SPEED TRAIN

The earliest high-speed train (HST) system is the Japanese Shinkansen, which dates back to 1964, running on the Tokaido route between Tokyo and Osaka. Later, this system was extended in various directions, of which one (the Joetsu Shinkansen) turned out not to be very successful (extremely high construction costs, record deficits followed by the privatization of the Japan Railway System) (see also Taniguchi, 1995a). In Europe, the more recent TGV and ICE are technically ready for large-scale application, and the first long-distance sections have been opened in the past few years. Both systems are currently being adapted in view of the realization of a larger European network. This planned HST network should cover 30 000-35 000 km, including about 20 000 kilometres of entirely new infrastructure. The related investments are estimated to be $235 billion (Ellwanger and Wilckens, 1993). In France, Germany, Spain and Italy several sections have already been opened, whereas in most other countries in Western Europe plans are nearly completed to start construction. An HST system involves fast trains stopping at just a few selected major stations on a long-distance route (Nijkamp *et al.*, 1994).

If the possibilities of this mode are to be used in an optimal way, it is necessary for a completely new infrastructure to be constructed. If, however, the demand is not sufficient, this would not be profitable. The solution in this case may be for an HST to use upgraded existing infrastructure, though this reduces the maximum speed. In the next section the success and failure factors of Chapter 4 will be applied as a frame of reference for judging the HST.

6.2.1 Technical factors

The most important HST systems are the French TGV, the German ICE and the Japanese Shinkansen; two less mature systems are the Italian Pendolino ETR450 and the British IC125. General characteristics of these systems are presented in Table 6.1.

Table 6.1. Characteristics of several HST systems

Country	France TGV-A	Germany ICE	Japan Shinkansen	Italy ETR450	UK IC125
Max. speed (km/h)	300-350	270	270-300	250	225
Max. gradability (per cent)	2.5	1.5			
Vehicle costs ($\$10^3$/seat)	35	45	20		

Sources: Hondius (1990); *Ingenieurskrant 23* (1993); *Technieuws 29* (1991)

As mentioned above, the Shinkansen is the oldest system. The TGV has been in operation since 1981. Present trains are not airtight so that high-speed operation in tunnels would lead to intolerable pressure drops. The newest trains built after 1996 are more airtight, however. A TGV train is able to draw 9 MW of power, and operates at a voltage of 25 000 V. At lower voltages, for instance,

1 500 V (in the Netherlands) or 3 000 V (in Belgium) existing electric power stations cannot deliver the electric power needed to run at maximum speed without slowing down other trains considerably. Running at lower voltages also leads to resistive losses in the wires and power electronics of the TGV train. When operated on the existing network, the speed is therefore limited to 200 km/h.

The ICE is of a more recent date (1991). It operates at 15 000 V. As the German landscape requires more tunnelling (on some tracks 30 per cent of the distance is travelled underground), this system has been airtight from the beginning. The low gradability, reported in Table 6.1, is actually not a special characteristic of this system, but a consequence of the requirement that the German high-speed rail network should also be able to carry the much heavier goods trains. In France the TGV-network is solely constructed for passenger transport.

The minimum radius of curves for the above systems at maximum speed is rather high (± 6000 m). This results from the fact that the cabins are fixed to the under-carriages. In the Italian Pendolino system this problem has been overcome by using tilting coaches. However, this involves a rather heavy, complicated and expensive construction, which may not be able to deal with train vibrations occurring at speeds above 300 km/h.

The TGV and ICE systems are technically ready for large-scale introduction, and present developments and policies are geared to the realization of a large European network. Both train systems are therefore being adapted to comply with railway systems in other European countries. In due course, the two systems - which were initially quite different - are expected to become more and more compatible. A basic European HST network may be ready between 2005 and 2010.

Eventually the HST systems may be able to reach maximum speeds of up to 500 km/h. Power transfer by the pantograph/catenary may eventually be the limiting factor in increasing speed (Lever, 1993). Also vehicle tilting will be necessary to improve the speed on curves. However, increasing the operating speed will cause excessive wear and increase life-cycle costs significantly.

The final energy consumption of HST is estimated to be 86 Wh per seat kilometre (Lever, 1993). Given an efficiency of 39 per cent for electricity generation, this figure agrees well with the primary energy consumption of 220 Wh per passenger kilometre as stated by McClintock (1993). At high speeds energy consumption is dominated by air drag; therefore improving aerodynamics is the best way to increase efficiency at high speeds.

Apart from energy consumption and related emissions, the environmental impact of HST includes noise, vibrations and segmentation of landscapes. In *Versuchsgesellschaft* (1985) the noise production from an HST at 250 km/h is estimated to be about 93 dB(A), measured at 25 m distance from the track. This noise nuisance may eventually be reduced by subterranean construction, which would also decrease the impact on landscapes, but this technology would entail high costs (see Section 6.5).

If the HST is operated on a designated infrastructure without level crossings, the system can be at least as safe as that of conventional trains.

6.2.2 Spatial factors

It is foreseeable that the adoption of HST technology will take place in a relatively large number of countries, provided that its spatial threshold level of demand is satisfied. If (inter)national urbanization patterns develop according to the chains and zones models discussed earlier, the interdependent metropolitan areas need to be sufficiently large (around a few million inhabitants) and the distances in between need to be sufficiently long to take advantage of the high speed. If future urbanization patterns develop according to the specialization and concentration model, the HST adoption seems realistic if there is a sufficiently large interdependency (and thus demand) between the top metropolitan centres in a country and between the interconnected international metropolitan centres.

A successful operation of HST systems at the interurban (regional) level also seems to be dependent on urbanization patterns on lower spatial scale levels. The fact that HST systems are not in use in the United States can be partly explained by the large sprawl in metropolitan areas of both homes and business activities (Hall and Banister, 1995). This indicates that a large-scale introduction of the technology does not seem plausible in a future model of diffuse cities.

6.2.3 Institutional factors

The HST can be characterized as an innovation in the existing collective transport system for national and international transport, and is increasingly acquiring an international character. At this level, horizontal and vertical coordination problems may form an important failure factor. With regard to vertical coordination, a distinction can be made between international coordination problems (EU versus Member States) and regional coordination problems (relationships between nation, region and local authority). The problems of horizontal coordination occur chiefly between the various countries involved in the project, since decision making may suffer from delays. This is the main reason why planning and construction of an HST may take 10 to 20 years.

Another problem may concern the organization of the operation by the national railway companies. These are all oriented towards their national market, and they are only responsible for the infrastructure and transport in their own country. This makes it difficult to operate an international section or network, which may reduce the advantages of the above-mentioned network synergy (see

Rühl, 1991). It is, for example, possible that certain sections will run at a loss, although they contribute to the profitability of other parts of the network.

There are plans to operate international sections by means of joint ventures of the railway companies, so that the exploitation is the responsibility of one organization. Whether these joint ventures are able to work efficiently is often doubted, because it is not likely that the individual companies would want to give up part of their independence and responsibilities. This could make decision-making a problem when an HST has to compete with other modes, for example, when the price of air tickets is lowered on a particular section.

The development of an HST is being strongly encouraged by national governments. The decision to develop it is mainly based on industrial policy criteria, which has led a number of European countries to develop their own system. However, in the development of a European transfrontier network the lack of standardization is now causing various problems. It is, for example, quite predictable that Germany will not accept the TGV standard, given the enormous amount of money that they have spent in developing ICE and the vast interests of their national railway industry. There is a tendency of national governments to protect national producers of infrastructure and vehicles (e.g. trains) by means of exclusive government contracts incorporating national standards. A compromise to overcome this problem consists of the adoption of a set of standards to which both systems should apply, to ensure the necessary compatibility, at least for transfrontier connections (Nijkamp and Vleugel, 1993).

6.2.4 Economic factors

Analysis of the HST market

The HST aims to serve the business market over medium to long distances. It is expected that HST can compete with other modes up to travelling times of 4.5 hours. This is the maximum time, which it is possible to travel both to and from a given destination in one day. To serve the business market HST will have to offer - next to a high average speed - a high level of comfort and service. In the rush hours (when there is most business travel), the price will be relatively high, probably higher than the charges of conventional trains nowadays (Commission of the EC, 1993b).

The HST should also aim at other markets to keep seat occupancy high. To achieve this, it should try to capture segments of the recreational market like day-trips and short holidays (e.g. 'Eurodisney'). To attract enough travellers the price will have to be much lower than during the rush hours. Such a lower price will be possible because the marginal costs of transport will be low, so that these costs would be covered by the revenues.

The new transport mode will attract new travellers, who did not travel on its route before, while there will also be substitution from existing modes. The HST will be a competitor of conventional trains in the first place, because existing services will be shut down or reduced, and will also compete with the private car because of the speed and comfort, especially on longer distances.

The expectation is that the HST will compete with the plane up to a distance of about 500 kilometres. The competitive advantage is that it connects city centres, while airports are often positioned quite far away from city centres. Another advantage is the quick check-in and check-out time: on medium distances this accounts for about 75 per cent of the total travelling time of air transport. All this means that transport to and from HST-nodes is much faster, so that the door-to-door travelling time is relatively short. It is clear that links with (relatively) poor air connections will benefit most from the introduction of the HST.

An example of the above-mentioned potential impacts is the planned HST-section between Amsterdam and Paris. It is estimated that transport by train between the Netherlands and France will increase by 75 per cent. Of this increase, 50 per cent is expected to be attracted from the private car, 18 per cent from planes and 32 per cent would be new travellers. The total volume of travellers will be 30 per cent higher compared to the period before the construction of the HST (Tweede Kamer, 1991).

There are also other possibilities of using the infrastructure, in order to make the exploitation more profitable. In Germany, for example, ICE sections are also used for the transportation of goods; in France TGV tracks are used for express goods and other post trains. Competition with air freight is also possible in this way. Other possibilities to generate revenues include providing catering services for travellers and renting/selling of commercial locations in and around stations.

The high-speed train as part of the transport network

Because the HST has just a few stops and a large capacity, it is important that the density of demand is high. Therefore substantial transport flows are essential. Of crucial importance is the quality of the transport to and from the nodes. When this is high, the potential market is much bigger, because the travelling time is attractive for more people. Good train connections with smaller cities and the existence of high-quality local and regional transport are important prerequisites.

An advantage of the HST is that the mode will be operated by the national railway companies in most cases, so that supplementary connections with other destinations may easily be arranged. The mode is also able to use existing stations, which are junctions of several transport modes. In this way it can participate in an existing transport network and may benefit from network synergy. This means that it does not have to create a completely new market; so exploitation might become profitable in a shorter time period.

The HST fits in well with the existing policy objectives of governments, because its negative external effects are much lower than those of competing modes like the private car and the plane. Furthermore, it is expected to be important for the structure and the competiveness of national economies; hence, governments are likely to support the introduction of this mode, which is an important success factor.

Financing HST

Until now some HST-sections have been financed by the private sector. These sections are the TGV *Sud-est* (Paris-Lyon) and the Channel tunnel (see Section 4.5.3). There are also several other plans for using private capital to (partly) finance sections, e.g. in the Netherlands. The problem is that the most profitable sections will be constructed soon, while for other sections this is less likely. In that case there may be no network synergy, which reduces the profitability of the total network. It is clear that the government will then have to finance a large part of the infrastructure, because the returns on investments are too low.

However, the exploitation of the HST infrastructure is expected to be profitable in most cases, if infrastructure costs are not taken into account. Especially in the European core zone, there are many large cities in a range up to 500 km, with significant traffic flows. In southern Europe the profitability of the network will be lower because of the lower population density and the lower purchasing power, which make traffic flows smaller.

The variable costs of exploitation are in general equal to those of conventional trains. Fixed costs, such as depreciation and maintenance costs, are about twice as high, but it is expected that the rolling-stock will travel twice as many kilometres, because of the high average speed. This is why there is no disadvantage in terms of higher fixed costs, compared to conventional trains, though with an HST it is important to keep the seat occupancy high.

6.2.5 Socio-psychological factors

The HST can be either seen as an improvement of the existing train system or as a new mode. The TGV *Sud-est*, for example, was marketed as an improvement of the existing train system, to increase the acceptance and adoption of the system.

When new infrastructure has to be constructed, the resistance of society will be high, especially when it cannot be linked to existing infrastructure. This problem may especially arise in rural areas, where the inhabitants have no direct benefit from the mode. This can be a reason to opt for more expensive infrastructure (tunnels, etc.) to restrict external costs and public resistance. It is then very likely that the initial cost estimates will be exceeded. The environmental benefits of the system should be weighed against the environmental costs. Nevertheless, in many cases acceptance might still be low even if the more environmentally friendly option is chosen.

If new infrastructure is not needed, however, public acceptance may well be high, because of the positive environmental effects, if the HST is a substitute for the car and the plane. Its adoption will depend on its attractiveness, which differs per section. When it is an attractive alternative for car and air traffic (for example, because it avoids a high level of congestion), it might be a success. This may also be the case because of the lower user costs (compared to a plane) and its high average speed (compared to a car).

6.3 PEOPLE MOVERS (MAGLEV LOW-SPEED)

There are numerous types of urban transport that are called 'people movers'. Some of them are already in operation, while most types are still at the R&D stage. Automatic rapid transit systems can be subdivided into Group Rapid Transit and Personal Rapid Transit (Makkink, 1994). Examples of Group Rapid Transit are the M-Bahn and the Monorail. In general, these are small driverless trains. Due to their high frequency they have capacities of up to 30 000 passengers per hour per direction, compared to about 60 000 for heavy rail systems. Headways of about 1 minute are feasible.

Around the world a small number of Group Rapid Transit systems are in operation. A well-known example is the VAL in Lille (France). Many of the current urban metro systems are to a large degree being operated automatically, but are not yet driverless.

Personal Rapid Transit systems can be described as track taxis, which have a maximum frequency of about six departures per minute. At present, no Personal Rapid Transit system is in operation; however, the construction of a test track near Chicago has recently been approved (Makkink, 1994).

A possible and attractive drive technology for people movers is the Maglev (magnetic levitation) technology. Therefore, in this section we discuss first the Maglev technology. Then we consider aspects of Maglev low-speed and high-speed trains, respectively.

6.3.1 General aspects of the Maglev technology

There are basically three different operating principles for high- and low-speed magnetic levitation (*Polytechnisch Weekblad 29/30*, 1993). First, a technically simple system - but impractical for large-scale application - is levitation with permanent magnets.

A second principle is called electromagnetic suspension (EMS). With this technology the vehicle is carried and guided by means of the attraction forces between conventional electromagnets in the vehicle and iron beams in the guideway. A problem with the attracting force is that it increases with decreasing distance between magnet and iron. Therefore, a rather complicated active control system is necessary to maintain a constant air gap of about 10 mm. In the EMS system the train is levitated at all speeds, so that this system can be applied to both low-speed and high-speed Maglev systems.

The third system is called electrodynamic suspension (EDS). It uses the repulsive force between superconducting coils in the vehicle and aluminum loops in the guideway for levitation and guidance. As the repulsive force increases with decreasing distance, the suspension and guidance in this system is self-regulating. Dynamic suspension only works at high speeds, so that at low speeds the vehicle will have to be supported by wheels. The EDS system is therefore not suitable for low-speed Maglevs. It usually has an air gap of about 100 mm and is therefore much more tolerant of irregularities in the guideway, and also less sensitive to external influences (weather conditions, earthquakes, etc.).

The superconducting magnets are operated at temperatures below 10 K, requiring refrigeration with liquid helium. A difficulty in using superconducting magnets is the possibility of quenching. When a superconducting magnet quenches, its wiring switches from the superconductive state into the normal, resistive state. In that state the magnet is unable to maintain a high current and the magnet will dissipate its magnetic energy. This energy is absorbed by the liquid helium which evaporates out of the cryostat. Quenching can be induced by vibrations or by large induction currents due to uncontrolled interactions with conducting objects (e.g. parts of the guideway structure).

Another disadvantage of the EDS system is the high stray fields inside the train. These may amount to some 10 Gauss, while the natural level is 0.5 Gauss; therefore shielding may be necessary.

Propulsion of Maglev trains is carried out by means of a linear induction engine, of which there are several types. Their operating principle is analogous to that of rotating electric engines, with the guideway acting as stator and the magnets on the vehicle acting as rotor. The linear motor has the following advantages compared to rotating electric engines (Van Zeevenhooven, 1993):

- it has no moving parts;
- it is independent of adhesion (wheel/rail friction), giving the Maglev systems high acceleration capabilities and a high independence of, for example, weather conditions;
- vehicle speed and position are controlled by the operating voltage and frequency of the stator field, providing a simple solution for automatic 'inherently safe' operation.

The linear motor has, however, the following disadvantages:

- its efficiency is lower than that of a rotating motor, e.g. because of stray fields at the poles of the magnets, larger air gaps and magnetization losses;
- the motor has to be dimensioned for the highest required power, as no overload is possible; as a result, rotating electric engines tend to weigh less.

In the next subsection, we discuss ways of using the Maglev technology for low-speed applications, while in Section 6.4 we elaborate on possible applications in high-speed trains.

6.3.2 Technical factors

Low-speed Maglev systems have been developed in Japan (HSST), the UK (Maglev People Mover) and in Germany (M-Bahn).

The Japanese HSST system is a Maglev vehicle without wheels which is supported and guided by attraction forces between electromagnets and iron rails (EMS), and propelled by a linear induction motor (Fujino, 1993). The development started in 1974, and the first vehicle was launched in 1989 at the Yokahama Expo. Presently, the latest HSST-100 system is being tested on a 1530 m test track with one switch. The maximum speed is 110 km/h, but speeds of up to 500 km/h are technically feasible and have been demonstrated experimentally. The vehicles are made from aluminum; the minimum radius of curvature at high speed is 100 m, while the energy consumption is 0.88 kW/tonne at a speed of 100 km/h. At this moment, this is the only system which is being developed further.

The technical characteristics of the UK and German systems are compared with the characteristics of competing conventional systems in Table 6.2.

The British Maglev People Mover has been developed for short-distance shuttle services (Van den Berg, 1993); the technology used is the above discussed EMS system. Since 1985 a shuttle service has been in operation between Birmingham Airport and the National Exhibition Centre. Recently, this service was discontinued, however.

The German M-Bahn system has been developed for short- to medium-distance shuttle services. The M-Bahn also uses magnetic suspension (EMS) and rubber wheels for lateral guidance. A test track was built in Berlin in 1989. The switches turned out to be a severe problem, however, because these appeared to be unreliable, noisy and uncomfortable. The test track was dismantled in 1991.

As is shown in Table 6.2, the energy consumption per passenger kilometre (100 per cent occupancy) of present low-speed Maglev systems is about 35 Wh. This is comparable to tram and metro, but 50 per cent less than for a conventional train. An important aspect of the EMS system is that it requires energy for suspension, regardless of whether the vehicles move or not. The power required for suspension amounts to some 1 to 2 kW/tonne.

The greenhouse gas emissions depend on the way electricity is generated (see Chapter 5). The noise production of low-speed Maglevs is less than that of a train or tram. However, the infrastructure is incompatible with existing systems.

Table 6.2. Technical characteristics of low-speed Maglev systems in comparison with tram, metro and train

	Maglev UK	M-Bahn Germany	tram	metro	train
Support	magnetic	magnetic	rail	rail	rail
Lateral guidance	magnetic	rubber	rail	rail	rail
Propulsion system	lin.mot.	lin.mot.	ac	ac	ac
Points / junctions	none	active	classic	classic	classic
Min. radius curv.					
* technical (m)	40	50	75	200	150
* comfort (m)	130	220	109	150	140
Top speed (km/h)	54	80	50	80	125
Oper.speed (km/h)	30 - 43	30 - 40	20	34	33
Vehicle capacity					
* seats	12	28	35	80	144
* passengers	40	80	80	220	280
Line capacity (10^3 pers./h/dir.)	1.6-3.2	3.2-9.6	4-8	6.6-26.4	8.3-33.4
System costs ($\$10^6$/km)	20	50	10	33	40
Operational costs ($\$10^3$/km/y)	575	1125	1375	1250	1800
Energy cons. (kWh/vehicle km)	1.2	3.0	3.2	7.1	18
Energy cons. (Wh/pass.km)	31	37	40	32	65

Source: Van den Berg (1993)

Elevated guideways can be used, which leave space for other activities and for intersections with other (transport) infrastructure. Switches are a complicating factor, as they involve flexible guideway sections which have to be positioned with high accuracy and reliability.

6.3.3 Spatial factors

Adoption of people movers is dependent upon a high demand for transport. On the scale of the metropolitan area (region) therefore, adoption of the people movers seems only realistic under conditions of a compact city with a high density demand. However, here there is a complication, because land for new infrastructure will not be readily available in densely populated areas. From a spatial point of view, therefore, people movers may only be adopted on a small scale, at least when the system is constructed as surface infrastructure.

6.3.4 Institutional factors

Because the mode will not be profitable and a lot of infrastructure is needed to have full coverage in already congested urban areas, government support is necessary for the technical development and introduction of this mode. Whether such a policy is sufficient to warrant large-scale application, still remains to be seen.

6.3.5 Economic factors

Analysis of the people mover market
Nowadays, passenger transportation within city regions of Western Europe is supplied by a wide range of competing transportation systems (see, for example, Van Witsen *et al.*, 1987). People movers are a much faster form of transportation than conventional bus and tram systems, which is their main advantage. The minimum station distance has to be about 400 metres, because otherwise people are faster on foot.

People movers are especially suitable for urban mass transit - that means for applications in the capacity range of 2 000 to 30 000 peak passengers per hour per direction (Emanuel, 1991). When trains are made up of several vehicles, train splitting strategies can be applied. This means that such systems can serve both the town centre, where high capacities are needed, and the suburbs, with lower peak demands.

Another market for people movers may be airports and business parks, which have a relatively high off-peak demand. The systems generally require little space and can move through buildings, thereby minimizing walking distances.

Financing people movers
People movers are not likely to become commercially profitable in the near future, but due to their favourable external effects (e.g. safety, emissions, congestion) they may be subsidized by the government. In the case of airports, business parks, recreation parks, etc., people movers may be used for the transportation of passengers to and from remote car parks. This means that quite

expensive space near, for example, airports (e.g. parking lots) can be used alternatively.

Both the track and the stations can be made significantly smaller than those of conventional modes. The infrastructure for people movers may therefore be up to 60 per cent cheaper than the conventional heavy rail infrastructure, when no subterranean construction is necessary (Emanuel, 1991). Compared to conventional services the operating costs are also considerably lower, which is mainly due to the fact that there is no need for a driver or attendant on board (Van den Berg, 1993). However, since the infrastructure is more expensive, the overall costs of people movers are comparable with those of bus services.

Compared with conventional heavy rail systems, the construction period is relatively short. Besides, since the space requirements are relatively modest and the noise levels are low, political decision making may be speeded up considerably. Because low-speed Maglev is a new technology, the financial risks are significant, however.

6.3.6 Socio-psychological factors

Acceptance of people movers in urban areas might be a problem, because of the lack of space in these areas. Where space is available, they have to compete with other modes or alternative uses of space. On the other hand, the introduction might reduce congestion in urban areas and emissions of pollutants, which would decrease smog. It is not yet clear whether the positive perceptions will outweigh the negative ones.

6.4 MAGLEV HIGH-SPEED

Recently some plans have been developed for the traction of Maglev high-speed trains. In Germany the construction of the first commercial project - the section Hamburg-Berlin (Transrapid 07) - has started (see also Wiescholek and Mayer, 1992). The estimated costs for this section (284 km) are DM 14.5 billion (NRC, 1994). There has been an experimental track in Emsland for several decades and in the United States a section is planned between Las Vegas and Los Angeles; the estimated costs of this project (427 km) are $6.25 billion (*Intermediair*, 1990). In these two countries there are other similar plans, but in other European countries there are no definite plans (Vranich, 1992). Another system is under development by Japanese Rail (McClintock, 1993; Lever, 1993). The profitable top speed of Maglev high-speed trains is about 500 km/h, compared to the top speed of HST of 300 km/h at present.

Maglev trains will drive without in-between stops, so the travelling time will be very short. By cutting out stops *en route* the higher speed of Maglev is used in an optimal way. Since the Maglev is a main competitor of HST, we will discuss its success and failure factors in relation to those of HST.

6.4.1 Technical factors

A description of the different operating principles of magnetic levitation has been given in Section 6.3.1. In Table 6.3 the various Maglev systems are compared with the French TGV system (see also Geerlings, 1995).

The acceleration capability of high-speed Maglev systems is two to three times greater than that of the TGV, so that in principle shorter trip times are possible. For distances of less than 800 km the travel time of Maglev systems is comparable to that of air transport. We will now discuss the German, Japanese and US application plans in more detail.

The development of the German Transrapid system started as early as 1968. The Transrapid uses EMS technology with a levitation system and guidance by the attractive magnetic force between electromagnets in the vehicle and iron beams in the guideway. It is propelled by a synchronous iron-cored linear longstator engine. It uses aluminum sandwich vehicles with 32 magnets per section. Guideway sections are individually energized when the vehicle enters the next section. The flux-coupled electromagnets on the vehicles are arranged in a redundant configuration and controlled in two degrees of freedom. On-board power is supplied by linear generators (induction) and batteries. Braking is provided mainly by Eddy-current brakes. The air gap is approximately 1 cm, which may be a problem for application in the US (earthquakes, unsteady soils, etc.), because it requires extremely stiff guideways (Lever, 1993).
Trains may consist of 10 units, with the possibility of mixing passenger and goods units. Passenger trains have 56 to 113 seats/unit; the maximum speed is between 400 and 500 km/h (Miller and Wackers, 1993; Raschbichler *et al.*, 1993; Versuchsgesellschaft, 1985).

Development of the Japanese Rail (JR) system started in 1970. In 1979 a speed of 517 km/h was achieved with a superconducting system. A new 43 km test track is under construction for testing a practical system for speeds of up to 500 km/h, which may be opened by the end of this century.

The JR-system uses EDS with superconducting magnets for levitation and a linear synchronous motor for propulsion. A null flux guidance principle is used, so that currents are only generated when the train moves out of its equilibrium position in the guideway. The levitation system is mounted in the side-wall of the guideway to improve stability and decrease energy consumption. At low speed the train moves on rubber wheels; the air gap is about 100 mm. Quench problems with superconducting magnets are not yet completely under control, however (Kaminishi *et al.*, 1993; Masada, 1993; Miyata *et al.*, 1993).

In the United States EDS systems have been designed by Bechtel and Foster-Miller. Grumman has designed an EMS system with superconducting coils and a 40 mm air gap. The Magneplane uses 'sheet levitation', also by means of superconducting magnets.

Table 6.3. Comparison of the characteristics of Maglev systems and TGV

	TGV	TR 07	Bechtel	Foster-Miller	Grum-man	Magne-plane
Low speed accel. ($g_{max.}$)	0.05	0.096	0.14	0.15	0.097	0.17
Min. radius of curves	6000	5800	2600	2800	3300	2200
Propulsion eff. at 480 (300) km/h (per cent)	(0.78)	0.83	0.87	0.91	0.78	0.84
Final energy consumption (Wh/pass. km) at 480 (300) km/h	(86)	125	165	95	100	125
Cost/seat ($\$10^3$)	41	58	39	93	71	200
Dual guideway costs ($\$10^6$/km)	14	10-12	14-17	9-20	8-10	13-15

Source: Lever (1993)

A comparison of the energy consumption of the distinct systems is presented in Table 6.4. In this case primary energy consumption is expressed in Wh per passenger kilometre, but the assumed occupancy factors are not given.

At 400 km/h the final energy consumption of Maglev systems is 58 to 64 Wh per seat kilometre; at 200 km/h it is 26 to 29 Wh per seat kilometre (Raschbichler *et al.*, 1993). In McClintock (1993) the dependence of the final energy consumption of Maglev systems on vehicle speed is discussed. The Maglev system has its lowest energy consumption at 180 km/h (65 Wh per passenger kilometre).

Table 6.4. Primary energy consumption and related CO_2 emissions of Maglev systems in comparison with other modes

	Primary energy consumption (Wh/pass.km)	CO_2 emission (g/pass.km)
US-Maglev	625	30
Transrapid 07	1100	N.A.
Transrapid 07 (8-seats)	470	N.A.
Plane	920	210
HST	220	10
Car	750	180
Advanced car	160	N.A.

Note: N.A. = not available
Source: McClintock (1993)

At lower speeds the energy consumption per passenger kilometre increases due to the constant power requirement for levitation. At 90 km/h final energy consumption equals 80 Wh per passenger kilometre. At 400 km/h final energy consumption is equal to 180 Wh per passenger kilometre. This means that the system can be operated at half its energy consumption by reducing operating speeds. From all this, the paradoxical conclusion is that if the Maglev system is to play a role as an energy efficient transport mode in a sustainable transport system it may not be operated at its highest possible speed. The CO_2 emissions depend in general on the efficiency and energy source mix of the electricity generation system.

As the Maglev is a contactless system there is no rolling noise, nor is there much engine noise, but aerodynamic noise becomes dominant above 200 km/h. At speeds between 100 and 200 km/h noise levels are around 70 dB(A) at 25 m distance, which is 10 dB(A) less than a car and 15 dB(A) less than a train at the same speed. At 250 km/h the noise level at 25 m is 82 dB(A), which is 10 dB(A) less than a train at the same speed. At 400 km/h the noise level equals 95 dB(A), i.e. a few dB(A) more than an HST at 250 km/h (Versuchsgesellschaft, 1985).

Due to the high gradability and the short radius of curvature, the guideway can be flexibly adapted to the landscape. This reduces landscape impairment compared to HST systems. The guideway structure of a Maglev system generally consists of steel or prestressed concrete beams on piers between 5 and 20 m high. The span between the piers is about 35 m.

For the Transrapid system the radius of curvature is 4000 m for speeds of about 400 km/h, 1000 m for a speed of 200 km/h and 600 m for speeds around 100 km/h. The space requirement per kilometre track is 50 per cent less than for a train (both double track) and 60 per cent less than for a six lane highway.

The safety of the Transrapid system has been proven through extensive testing. The system has several inherent safety features. First of all, there is no possibility of derailment in case of accidents as the vehicle embraces the guideway. Secondly, the vehicle speed is automatically controlled by means of the linear motor drive, so there is less chance of vehicle collisions. Maglev trains in general use dedicated infrastructure, so that there is no interference/level crossings with other transport systems. Overall safety may therefore be better than that of train systems.

6.4.2 Spatial factors

Two major negative spatial features may be distinguished in the spatial field: the need for a new infrastructure and a high threshold level of demand. Furthermore, minor negative features are concerned with local impacts, i.e. aerodynamic noise produced at high speed. This noise may be incompatible with spatial functions such as living quarters and nature reserve areas, and makes necessary a careful (and potentially expensive) selection of the route of the system. Currently, however, studies are being made to find an optimum vehicle shape and to solve the noise problem (Taniguchi, 1995b). There is also the negative feature of segmentation of landscapes, but due to a high gradability and short radius of curvature, the guideway can be flexibly adapted to the landscape. This reduces landscape impairment compared to HST.

A problem may be its high spatial threshold level of demand. When national urbanization patterns develop according to the chains and zones model (see Chapter 2), the interdependent metropolitan centres need to be sufficiently large and the distance needs to be sufficiently long to take advantage of the high speed. In this model, adoption may be hindered when the corridors between the metropolitan centres lack easily available land for new infrastructure. When future urbanization patterns develop according to the specialization and concentration model, adoption seems only realistic when there is a sufficiently large interdependency between the top metropolitan centres of a country.

6.4.3 Institutional factors

Maglev high-speed trains do not seem very competitive with HST in many cases, though there might be some pressure from national industries to introduce the mode, as was the case in Germany and Japan. Support by governments may then be achieved, which seems to be a necessary condition for the success of this mode. For the time being, it is mainly the private industry which, together with railway operators and national and regional bodies, has to overcome the institutional and planning barriers.

6.4.4 Economic factors

Analysis of the market for Maglev high-speed trains

Maglev is a transport mode with a top speed falling between the existing HST and the plane. The potential market, however, will be about the same as that for the HST, with the difference that Maglev might be competitive up to distances of 850 km (HST 500 km) with planes. In the first instance, Maglev will aim at the business market, and then at other submarkets.

It is clear that on most sections a choice should be made between an HST and a Maglev, because it is not conceivable to build two systems in parallel. The choice between the modes will depend on the characteristics of the mode. Therefore, this section is mainly concerned with a comparison between the two.

A disadvantage of Maglev is the higher infrastructure costs. These are US$11 to US$63 million a mile, whereas HST costs US$8 to US$32.2 million a mile depending on soil conditions, space costs, the number and steepness of slopes, etc. (TRB, 1991). It is possible that the construction of Maglev infrastructure will become cheaper in the future, because there will be more experience and returns to scale when more sections have been constructed (following a normal learning curve). On the other hand, it is also possible that it may become more expensive than is estimated now, because no commercial tracks have yet been constructed. Another problem may be that Maglev cannot easily be connected with the existing rail network, as is possible with an HST. This means that in several cases more expensive and new infrastructure is necessary.

On short sections (up to 80 km) it seems possible that a Maglev can compete with an HST, because a Maglev can accelerate and decelerate much faster. Plans for such short-distance tracks have been made in the US and Germany for sections from airports to city centres. It does not seem likely that a Maglev train would be very profitable on such distances, but they may be attractive in specific cases (Vranich, 1992). For instance, they offer considerable advantages for airports by increasing accessibility of the airports, which makes their use even more attractive.

Over medium distances HST is considered to be more attractive, because of the lower construction costs. These are considered to be more important than the higher speed of Maglev. Only on tracks where traffic in the air suffers from congestion, like the section from Las Vegas to Los Angeles, are the higher construction costs compensated by the higher speed. Also it is possible that Maglev can compete with HST in hilly areas, because Maglev can take steeper slopes (less tunnels, etc.).

On longer sections air traffic is probably too competitive. It may thus be concluded that over short distances and on sections which suffer from congestion in the air, Maglev may have a promising market potential (Brand, 1987; Breimeier, 1993; Miller and Wackers, 1993; TRB, 1991).

Maglev high-speed as part of the transport network

A big disadvantage of Maglev is that it is more difficult to integrate it in the existing rail network than HST, because of the different technology used. High-speed trains can make use of existing railway stations and infrastructure, while Maglev can only use completely new infrastructure. Diversions, for example, are not possible, nor is it possible to drive (partly) on existing tracks during construction or maintenance. Maglev may thus benefit less from network synergy.

Another problem is the transport of travellers to and from the stations, when completely new stations have to be built. High-speed trains will use existing railway stations in most cases, which are centres of all kinds of transport, like conventional trains, local and regional transport by trams, buses, etc. This is a significant disadvantage for Maglev. There is a possibility that Maglev could use existing stations; this will be relatively expensive, however, because all crossings with rail tracks would need to be constructed at separate levels. Another problem appears if both modes are operated by different private actors.

A conclusion may be that Maglev is probably more attractive in areas where not much rail infrastructure is available (e.g. in the United States) than in areas which already have a great deal of such infrastructure (e.g. in Europe).

Financing Maglev high-speed

The financing problems of Maglev are comparable to those of HST. It is likely that most sections will not be profitable, so that the government will have to (partly) finance the infrastructure. On balance, it remains uncertain whether the more expensive Maglev has sufficient advantages to make it preferable to HST in the majority of cases.

An advantage may be that as in the case of cross-border connections Maglev is a completely new technology, there will not be many standardization problems. If exploitation is not carried out by existing bureaucratic rail companies, the management of Maglev might be much more market oriented.

There is a big risk, however, that the costs will be much higher than expected, because the technology is still new and there is not enough experience with it yet. This makes Maglev projects risky to finance.

6.4.5 Socio-psychological factors

Because new infrastructure is needed for building a Maglev system, resistance in society may be high, especially if it is not linked to other infrastructures. On the other hand, the positive environmental impacts (compared to car and plane traffic) may increase the acceptance of the system.

Regarding the social adoption of Maglev, in general the same arguments hold as for HST. When there is heavy congestion in the air and on the road, the system may be successful, but this will of course depend on the particular characteristics of specific geographical sections.

6.5 SUBTERRANEAN TRANSPORT SYSTEMS

Subterranean infrastructure is an option which may in principle have several interesting applications. In fact, subterranean infrastructure is not a new infrastructure concept, as the first tunnel (in Boston) was constructed in 1850, while in cities like London and New York, for example, local subterranean systems were in operation before 1870. So, for local transport, a subterranean infrastructure is not new. Moreover, nowadays, there is quite a heavy flow of goods transport (e.g. oil, gas) using subterranean pipelines.

What is new, however, is the idea of constructing a subterranean infrastructure for medium- and long-distance transport by conventional modes. There is a clear trend in this direction: examples are the Channel tunnel and the Scan link connections in the Nordic countries. In Switzerland plans have been made for a network of subterranean high-speed trains through the whole country (Swiss Metro).

In general, subterranean construction of conventional modes is not likely to reduce emissions of greenhouse gases. Nevertheless, it does offer opportunities for diminishing other important environmental impacts of transport, such as noise and segmentation of landscapes. It may also serve to alleviate problems in terms of occupation of scarce space, congestion, etc.

Another and interesting possibility for subterranean transport is the development of entirely new transport modes. These modes are still in the R & D stage, but two derailed plans have been developed in the past few years. The Swiss Metro Project is intended to connect the major Swiss cities with a high-speed metro system (see also Marchetti, 1994). The system consists of two parallel tunnels, each with a small diameter (3.8 m) at a depth of 50 to 100 m. Vacuum technology and Maglev vehicle principles may lead to a maximum speed of 400 km/h. The Swiss system is only designed for passenger transport. The limited focus of the Swiss Metro system (largest cities, passengers) forms the basis for the resistance to the project. By far the largest transport problem in Switzerland is concerned with the transit of goods, but Swiss Metro offers no solution to this problem. A further but minor disadvantage in the Swiss case originates from the disposal of the removed rock debris (see also Bovy, 1994).

In the Netherlands, the concept of the High-speed Tunnel Transport System (HSTT) has been developed and advocated by private industries. This includes a network of tunnels in which a bullet-shaped vehicle is propelled by a linear motor (ISDS, 1988; 1989a-c). The HSST system is designed for both passengers and goods transport, including goods collection and distribution using containers.

The Dutch and Swiss designs of subterranean systems have to use various highly specialized technologies. Further R & D may help to avoid the danger that the resulting national systems may end up being incompatible. Their particular relevance lies in transnational transport, for example, between Europoort (Rotterdam) and the Ruhr area or Antwerp (Belgium). Thus, with a view to achieving a smooth interconnectivity, a sufficient compatibility of the hardware and software of the various national systems must be guaranteed.

6.5.1 Technical factors

There are several technologies available for building tunnels. Applicability and cost-effectiveness of these technologies will in general largely depend on tunnel characteristics and soil conditions (*Ingenieurskrant 20*, 1990; Stuurgroep, 1993). The basic tunnel parameters are diameter, depth and length. For tunnels just below the surface excavation is obviously the simplest technology. Disadvantages of the excavation technique are the fact that existing infrastructure (buildings, roads, etc.) positioned over the line of the tunnel will have to be demolished and the associated building activities will inevitably cause damage to the landscape.

For deeper tunnels various drilling technologies have been developed and demonstrated in Japan as well as in Europe. Drilling has the advantage that it does not disturb activities and infrastructure at the surface, and landscape damage is limited. Only for entrances, air-shafts, and for the supply of building materials during construction, do connections to the surface have to be made. Drilling requires special high-tech equipment and is therefore quite expensive. This technology can also be applied in less stable soils (*Ingenieurskrant 20*, 1990).

Subterranean rail networks

Subterranean operation of rail networks is not expected to cause improvements in terms of energy consumption. The main environmental objectives will be noise reduction and reduction of landscape segmentation.

Electric trains can be operated quite easily in tunnels, as they do not produce exhaust emissions. Tunnel ventilation can therefore be limited to a level required for the oxygen supply of passengers. For long-distance subterranean operation, trains will require some adaptations. A train moving through a tunnel at high speed causes pressure waves which are extremely uncomfortable to passengers. Therefore, train compartments will have to be airtight. This in turn will require adaptation of the compartments' ventilation system. Also the noise insulation of trains will have to be improved considerably (especially in the French TGV).

Subterranean road networks

Subterranean operation of a road network is more complicated than for a railway network, due to the fact that cars produce toxic exhaust emissions, tunnels will therefore have to be effectively ventilated to protect the health of car users. Ventilation shafts would in turn emit large amounts of pollutants, which may be undesirable in view of the local air quality at the surface, so high chimneys and/or air-purification equipment may be necessary. Also safety aspects related to the use of inflammable fuels, and collision risks resulting from individual vehicle control, are complicating factors at present. Collision risks, however, may be minimized in the future by introducing centralized vehicle guiding technology. Other hazards related to exhaust emissions and inflammable fuels may eventually be overcome by the use of electric vehicles or roadway powering (see Section 5.5).

Subterranean vacuum tunnels

As yet, vacuum tunnel transport systems only exist as a concept. One of the examples is the HSTT designed in the Netherlands. Due to the reduced air pressure inside the tunnels the vehicle experiences an extremely low air drag. The HSTT can reach a top speed of 500 km/h in only 5 km (ISDS, 1988; 1989a-c). The energy consumption is estimated at 66 Wh/tonne/km. A system capacity of 200 vehicles per hour is claimed. The HSTT system is fully automated.

The vehicle of the HSTT has a cylindrical shape, is placed on two air-suspended under-carriages and the vehicle has a pressure cabin which is under atmospheric pressure. Brake energy is regenerated, stored in the vehicle and used for subsequent acceleration.

6.5.2 Spatial factors

The major positive critical spatial feature of subterranean systems is their low occupation of scarce space. Land is only used for entrance and exit, stations and for air-conditioning. Future tunnels are different from conventional types with respect to the distances they bridge. Investment costs of future tunnels may be very high, so that the use of the systems are restricted to densely populated areas and corridors for large goods transport. From a spatial perspective, therefore, the major negative critical feature seems to be a high minimum level of demand for transport. As a result, subterranean systems will be feasible on the interurban (national) and international level when the trajectories include a considerable number of large and strongly interdependent population and industrial centres. In addition, the presence of natural barriers, including water, mountains and valuable natural reserve areas, may justify long-distance tunnelling (e.g. the Channel tunnel).

6.5.3 Institutional factors

Because of the high costs and the high (financial) risks, it seems almost obligatory to have strong government support in order to implement a subterranean transport system. To introduce transport through vacuum tunnels the pressure of national industries may be an important institutional factor. These industries may have been the prime movers behind the research concerning the proposed plans for the Swiss and Dutch systems. A major advantage is of course the low landscape intrusions and environmental decay, so that institutional barriers may be relatively less severe.

6.5.4 Economic factors

Subterranean infrastructure for conventional modes
The most important problems of subterranean transport are the high costs. Another problem is that the technology is quite new and hence will face many unexpected problems, especially in the Netherlands, where the subsoil is relatively soft. Therefore, substantial financial outlays may be lower in other countries. Nevertheless, it is quite possible that constructing subterranean infrastructure will become cheaper in the future, because of new technologies and learning-by-doing effects.

Sometimes, a subterranean infrastructure may be necessary, because it may not be possible to construct infrastructure above the ground owing to lack of space. In theory, subterranean construction is only profitable when the higher costs are compensated by the savings made (e.g. no costs for acquiring land, avoidance of expensive infrastructure like crossings at different levels, or energy savings). The first areas which are considered appropriate for the construction of subterranean infrastructure are urban areas, where space is scarce, which makes the costs of surface infrastructure high; these areas are also riddled with other infrastructure, so that many (expensive) crossings at different levels are needed, which adds to the costs. A Dutch study (Stuurgroep, 1993) reports that the costs of subterranean construction in urban areas are on average 30 per cent higher than those of surface infrastructure. In rural areas these costs might be 500 to 600 per cent higher. Subterranean infrastructure has one important advantage, however; the tunnels can be dug in a straight line, which makes the track shorter (and hence cheaper), and the average travelling time may then be shorter.

Recently, in the Netherlands a number of plans have been launched to construct the new Betuwe freight railway track to Germany (partly or entirely) subterranean. This track (>100 km) should connect the Rotterdam port with its main hinterland of Germany. According to ISDS (1993) the costs of constructing this line underground would be US$2.9 to US$3.3 billion, which is about the same as surface construction. Other experts, however, estimated the costs of these plans to be US$7.5 to US$9.5 billion, so that it is not clear what the real costs will be, but it is very likely that the costs will be higher than those of surface construction. Recently, the government decided not to construct the track fully underground but to use only local subterranean solutions (long tunnels) where there are serious environmental obstacles.

The lower external costs is the biggest advantage of subterranean solutions compared to surface infrastructure. Such a reduction can be achieved by less defilement of landscapes, and the avoidance of noise, stench and visual nuisance. Therefore, the government may decide to support subterranean construction of infrastructure, not only in urban areas, but also, for example, in areas of outstanding natural beauty. The government will then probably have to bear the higher costs, because there are no extra profits to compensate for them.

Market analysis of the High-Speed Tunnel Transport System

In Perrels and Schippers (1991) some first tentative calculations have been made about the market potential and the profitability of HSTT. It is expected that this mode may be attractive for distances between 150 and 1000 km. In this range, it may compete in the passenger market with HST, Maglev and plane. Because of the longer expected check-in procedures compared to train systems, the mode will probably not be able to compete with HST and Maglev, however, but it will compete well with planes on distances above 500 km.

Calculations of the market potential and profitability have been made for two routes: these are Amsterdam (Schiphol airport)-Munich and Amsterdam (Schiphol airport)-Paris and have the highest density in Europe. The cost estimates of construction for the first track are US$14.3 billion; the exploitation costs are expected to be US$1.3 billion and the revenues US$1.4 to US$1.6 billion a year; so in these calculations the mode might be profitable. It is clear, however, that these calculations are made on the basis of many assumptions. It is, for example, very likely that the infrastructure costs are underestimated, as has been the case in many previous major infrastructure projects.

Advantages of this mode are the price of the ticket, which is said to be lower than the airplane fare, the large capacity and the high speed. Other advantages are found in the external effects. The space costs are low (except around stations), the air pollution is much lower, the system is relatively safe and there is little noise, stench and visual nuisance. This may make the system attractive, even when it is not entirely profitable. So the government might well support the construction of this mode.

6.5.5 Socio-psychological factors

Because of the positive (or less negative) impacts of subterranean transport on the environment, acceptance will probably be high, while resistance in society will be small because of the reduction of various types of nuisance caused by this type of transport. This means that long procedures and delays may be avoided, which might save many costs.

The adoption rate might be low at the beginning, however, for psychological reasons. Potential users may especially fear travelling through vacuum tunnels, but this could be overcome by clever design and adequate marketing.

6.6 CONCLUSIONS

It appears that there are many technological options which may reduce the externalities caused by existing transport. The successful introduction of these options appears, however, to be dependent on many success and failure factors (see also Rienstra *et al.*, 1996). The main modes discussed in Chapters 5 and 6 have been summarized in Table 6.5.

At present, improvements of the conventional car, the HST and telematics systems seem to have the best chance of being introduced in the next decades. These do not require investments in alternative transport networks, and therefore they can be introduced gradually and implemented in existing transport networks. Despite this, their introduction already faces problems in many fields. But even when these options are introduced, they will not be sufficient to reduce the externalities of transport in order to make it more sustainable. This is because the advantages are counterbalanced by rising mobility and the expected minor improvements of the car and plane. Therefore, also (some of) the other options should be considered. Every option, however, faces several problems and may require significant changes in both government policies and individual behaviour.

To analyse such changes in a complex and uncertain environment, scenarios may be an interesting tool in which expert views may also be incorporated. Therefore, in Part C of this book we will construct various strategic scenarios for the future of the transport system, which take into account the success and failure factors discussed above.

Table 6.5. Tentative scores of new technology options[1)]

	Alternative fuels	Electric car	People Mover	HST	Maglev high-speed	Telematics systems	Subterranean conventional	Subterranean HSTT
Price/quality ratio compared to								
-conventional car short distance	-/+	-/+	-/+	--	--	+	-/+	-
-idem long distance	-	-	--	+	-/+	+	-/+	-/+
-airplane	--	--	--	+	+	+	-	-/+
Proven technology	-/+	-/+	++	++	-/+	+	-	--
Compatible with existing								
-infrastructure	n.a.	+	--	+	--	++	+	--
-fuel storage and distribution	-	-	+	+	+	n.a.	+	-
CO_2 emissions	-/+	$+^2/++^3$	++	++	++	-/+	-/+	++
Noise annoyance	-/+	++	++	-/+	+	-/+	++	++
Extra space needed	-	--	-	--	--	+	++	++
Social acceptance	-/+	++	+	-/+	-/+	++	++	-/+
Adoption	-	-	+	+	+	++	-/+	-/+
Expected government support	-/+	+	-/+	++	-/+	+	-	-/+

Notes:
(1) ++ very positive score + positive score -/+ no clear impact - negative score -- very negative score
(2) Produced with conventional fuels
(3) Produced with alternative methods

Part C

SCENARIOS FOR SUSTAINABLE TRANSPORT SYSTEMS

7 SCENARIOS FOR FUTURE SUSTAINABLE TRANSPORT SYSTEMS

7.1 INTRODUCTION

The need for more sustainable transport systems is becoming manifest all over the world. Options for coping with the environmental decay caused by the transport sector are manifold. In general, one may argue that externalities caused by transport may be reduced by means of four options:

- improvement (technical, managerial) of vehicles;
- introduction of more sustainable transport technologies and of new transport modes;
- a modal shift towards cleaner (mostly collective) transport modes;
- a reduction of the total mobility level.

Not all these options have the same chance of being adopted in society at large. The first option will normally lead to the least resistance in society and to the lowest necessary changes in individual transport behaviour. Nevertheless, as discussed in Chapter 5, current expected improvements in the private car and aviation technology are not expected to reduce the externalities to such an extent that a 'sustainable' transport system is easily in reach, mainly because improvements are simultaneously counterbalanced by growing mobility rates.

Therefore, alternative options - also of a non-technical nature - have to be considered as well. As demonstrated in Part B, there are also many new technical developments for future transport systems. It is clear, however, that the success of these innovations depends critically on many societal developments and policies, while it is not at all sure that they will solve all problems caused by transport. Therefore, it is meaningful to investigate a blend of all options in order to attain a more sustainable transport system and to analyse how far new technologies may be introduced as complementary and supporting mechanisms. In practice, sets of various mixed options may be applied, which will have different impacts on many - also non-environmental - fields. Therefore, trade-offs have to be made in order to identify sustainable solutions in the transport sector.

Scenarios are an appropriate instrument for analysing such unstructured policy problems with uncertain outcomes. Therefore, in Part C of this study we will construct various scenarios for the future of transport, which take into account the four basic spatial, institutional, economic and socio-psychological factors, as well as the technological potential needed for various sustainability options in transport systems. In this chapter we will first present an introduction to the scenario methodology.

7.2 SCENARIO FEATURES

7.2.1 General contents

Scenarios were used for the first time in the 1950s for US military defence planning. The basic definition of a scenario has been formulated by Kahn and Wiener (1967) as: 'a hypothetical sequence of events constructed for the purpose of focusing attention on causal processes and decision points'. In this respect scenarios are based on a vision, and do not necessarily need to be realistic. Scenario writing is 'a process to create, register, discuss, analyse, synthesize, document, store, retrieve and present information that is useful for the future and the development process to reach it' (Svidén, 1989).

In the context of scenario design, much attention should be given to the roles of various stakeholders, policy agencies or institutions in a dynamic world. Scenarios can assist decision makers in several ways (Taylor, 1990):

- identification of crisis signals by scanning the external environment;
- selection of stable policy solutions which produce robust results under varying conditions;
- assessment of emergency strategies to cope with threats from the outer world;
- risk assessment of various uncertain future developments;
- emphasis on an open mind and flexibility in decision strategies.

A scenario should normally consist of three parts (Nijkamp and Blaas, 1994):

- a description of the present situation;
- a description of a number of future situations;
- a description of a number of events that may connect the present situation with future ones (the path).

Every scenario analysis should ideally consist of and describe these parts. It is clear, however, that the emphasis may differ according to the particular scenario being developed. If there are, for example, already well-known descriptions of the present situation available, the emphasis may be laid on the other parts, while in other cases more attention should be paid to the present situation.

Using and constructing scenarios may provide new insights into possible paths and policies and their impacts on the future situation, and societal and political choices may become clearer. They provide a frame of reference for judgements, and help to identify strong and weak points in options addressing future problems.

It should always be acknowledged, however, that scenarios are not supposed to present an accurate picture of the future, but are designed to challenge the imagination, to provide new ideas and to present possible impacts of policy options, social developments and individual behaviour. In a decision process, the role of scenarios is akin to that of flight simulators.

7.2.2 Scenarios versus conventional methods of policy analysis

Scenarios differ from conventional methods of policy analysis and strategic planning *inter alia* in that uncertain roads to reach a situation are also explored. There are, for example, econometric forecasting models which predict exact figures of economic growth, inflation, employment rates, etc. Scenarios however, have a different scope and objective; their features and differences are summarized in Table 7.1.

Table 7.1. Advantages of using scenarios

From:	To:
- Focus on quantified variables	- Focus on qualitative pictures
- More emphasis on details	- More emphasis on trends
- Results determined by status quo	- Results based on future images
- From present to future	- From future to present
- Deterministic analysis	- Creative thinking
- Closed future	- Open future
- Statistical-econometric tests	- Plausible reasoning
- From simple to complex	- From complex to simple
- From quantitative to qualitative	- From qualitative to quantitative
- Single track thinking	- Multi-track thinking
- Reactive problem driven	- Proactive vision driven
- Multiple implicit assumptions	- Transparent simple assumptions
- Limited set of options	- Open range of options
- Model-determined mind	- Alertness to signals of uncertainty

Source: Zwier *et al.* (1994)

To achieve the above-mentioned advantages policy-relevant scenarios should contain:

- a variety of perspectives, which are not necessarily congruent;
- openness to envisage possibilities beyond current beliefs, plans and systems;
- coverage of crucial dimensions, in particular critical threshold conditions;
- assessment of different structures behind different images of the future;
- pathways or crossroads to different futures;
- a basis for early warning and proactive policy analysis.

A main difference with economic forecasts is that scenarios are not just extrapolations of current trends, but offer a set of different pictures which describe futures in a consistent way. In this respect, forecasting models have limitations when dealing with longer time perspectives, especially in an uncertain and unstable environment, because their aim is to predict the future.

7.3 SCENARIO TYPES

Scenarios can be designed on the basis of different principles. At least three different approaches can be distinguished:

- A **single technological paradigm** (e.g. magnetic levitation); this is a straightforward approach, in which a new technology is assumed as the core of a transport system, and then the consequences can be traced by a qualitative impact assessment (see Grübler and Nakićenović, 1991a).
- A **compound transport/land use package** approach (e.g. a transport system for zero-emission cities); in this case, a comprehensive set of relevant features of the transport system is combined with contextual factors (such as land use, commuting, behaviour, residential policy) in order to create a multifaceted image of a comprehensive nature (see Masser *et al.*, 1992).
- A **compound behavioural framing approach** in which the background conditions of transport behaviour (spatial, institutional, economic, socio-psychological) are mapped out (e.g. a deregulated urban transport system); in this approach the framing conditions are used to identify which transport technologies are in agreement with behavioural preferences and expectations, based, for example, on expert opinion. This approach is adopted in this study.

There are several ways of constructing scenarios: they may be intuitive, a literary product (mostly historic), idealistic, qualitative expert assessments or a starting-point for quantitative forecasts (Svidén, 1989). Most scenarios have various traits in common, such as the identification of risky policy areas, the assessment of critical threshold values, the recognition of different futures as a result of conflicting policy views, the need for balanced robust decision strategies or the design of creative or even path-breaking policy strategies. Several distinctions leading to different classes of scenarios can be made (Nijkamp and Blaas, 1994; Dreborg and Steen, 1994):

- a scenario may be either descriptive or normative.
- a scenario may be projective or prospective;
- a scenario may be commonsense oriented or expert-based;
- a scenario may be trend, reference or knowledge based.

Next, we will briefly elaborate the differences between these four categories of scenarios.

7.3.1 Descriptive vs. normative scenarios

Descriptive scenarios are mainly based on know-how about past and current trends and may also be based on expectations of how these trends will develop in the future. In this way no value judgements are included - apart from conventional wisdom, and major shocks or major policy changes are not assumed.

Normative scenarios include questions about the desirability of a development or choice. Normative scenarios may be constructed by applying norms and values of the authors (or, hypothetically, of the society), but also by means of surveying and asking people to provide normative information. Then it is possible not only to use majority responses, but also opinions of small(er) minority groups. In this way several - sometimes contrasting - scenarios may be constructed.

7.3.2 Projective vs. prospective scenarios

Projective scenarios are based on (qualitative) forecasting in which the future picture is based upon the current situation and future paths leading to it (Schwarz, 1988; Steen, 1994). Next, the impacts of those trends are described, which is the future picture. Also shocks and alternative developments may be assumed, but the path to the future always starts with the current situation. The main disadvantage of this approach may be that it is rather conservative, since it gives limited scope for the imagination. These scenarios are mostly descriptive in nature, and usually no value judgement is made.

Prospective scenarios are based on backcasting. Because we mainly use this approach in the present study, we will elaborate on it in more detail. In this approach first the situation in the future reference year is described, while next the paths (e.g. policy measures, societal changes) leading to this situation are presented. In this way it is possible to use the imagination of the designers to a large extent, and the process is more open-minded. The future picture may be constructed by, for example, external opinions ('desired situation'), investigating visions of the future dreamed up by, say, different political parties (Conservative, Liberal, Social-Democratic, Communist), or by assuming some (extreme) societal or technical developments. It is possible to make extreme scenarios based on assumptions which are not likely to become true, but it is also possible to construct more realistic scenarios.

Features of backcasting scenarios are (Robinson, 1990):

- they are normally explicitly normative;
- they determine the physical and social feasibility of that future;
- they analyse what policy measures are required to reach a future situation.

A necessary condition is that the time frame should cover a sufficiently long period, so that major changes are possible.

A disadvantage of the backcasting approach may be that the results are accepted to a lesser extent by politicians or society if they are incompatible with dominant views and current plans. This drawback is, however, inherent in this methodology.

7.3.3 Commonsense oriented vs. expert-based scenarios

Commonsense oriented scenarios take for granted existing views and opinions shared by the majority of the population and try to build new futures from such views and opinions. The danger is, of course, that such scenarios lead to common wisdom images without much imagination; the advantage is that they may easily enjoy much public support.

Expert-based scenarios use views from experts in a certain field, who are assumed to be able to assess realistic and creative pictures of the future without being constrained by current wisdom. They may also use desirable images and combine these with their expert opinion. This approach will be adopted in the present study, mainly combined with a backcasting approach.

7.3.4 Trend, reference and knowledge scenarios

Trend scenarios may describe the impacts and the final picture of current policies and trends. They may serve as a 'business as usual scenario' and in this way form the analysis on which the other scenarios are based (e.g. by assuming a 'shock'). In this way it may be analysed to what extent and where changes are necessary to achieve alternative futures and current goals. Trend scenarios may be constructed by using an intuitive analysis (e.g. by analysing current trends) or by surveying experts or others; they are descriptive and based on forecasting in nature.

Reference scenarios, on the other hand, may provide extreme pictures of the future, by analysing and assuming extreme developments and policies. In this way a future reference framework may be provided for less extreme scenarios, and they may also make policy choices and their impacts very clear. Reference scenarios should mainly be constructed by means of an intuitive analysis and may serve as a starting-point against which to compare less extreme scenarios, for example, the impact of several less extreme policy packages.

Knowledge scenarios may be constructed by using scientific insights, since it may be assumed that scientists may have the best insight into future developments, into possible, likely and desirable policy measures and into their impacts. These scenarios may have a higher credibility than if they are based on the views of just a few authors. Such opinions may be gathered in several ways; e.g. surveying, scanning of literature, attending seminars, organizing workshops, etc. Delphi methods may be especially attractive for constructing such scenarios (Svidén, 1989).

By constructing knowledge scenarios a distinction may be made between packaging and non-packaging. In the first approach consistent policy packages are presented to the experts, who may choose a desired or expected one; next they should indicate possible impacts (see Masser *et al.*, 1992). The advantage is that in this way a number of consistent policy scenarios are constructed, and then the impacts of several policy packages are presented. A disadvantage may be that alternative - potentially more creative - ideas are not analysed, which limits the scope of the scenario experiment.

The second approach is non-packaging (or menu-driven) in which the experts may choose their own blend of policies. In this way many new ideas may be gathered, e.g. by constructing minority scenarios by combining different groups from one sample; this also offers the opportunity to investigate in more detail possible inconsistencies in opinion among different segments of one sample. A disadvantage of this approach is that the packages chosen may not be entirely consistent, although the experts may be expected to develop them more or less consistently.

7.4 METHODOLOGY ADOPTED IN THE STUDY

In light of the great many uncertainties described in the previous chapters, the choice for a scenario analysis seems plausible. The main question is whether it is possible to identify relevant and interesting future images which may be used as a frame of reference for developing and evaluating sustainable transport strategies. This means that the scenarios designed should be *policy-oriented* (for both the private and the public sector), seeking transport-related strategies to reduce the negative externalities of transport. Consequently, various options have to be investigated, such as technical improvements, modal split changes, improvement of transport efficiency through telematics, new ways of physical planning, or changes in behaviour or lifestyle.

For a proper judgement of scenarios, a solid *anchor* (or *frame of reference*) is needed. Sometimes the status quo acts as such an orientation point, but this may sometimes constrain the creativity in scenario design. In our scenario experiments, we have adopted two extreme reference scenarios - each characterized in eight distinct dimensions - which provide a reference pattern.

Any other scenario to be designed will fall in between these extremes, so that it is always possible to evaluate such scenarios on the basis of a difference analysis with respect to the reference scenarios.

Another feature of our scenarios will be that they are explicitly *bottom-up oriented*. This means that the scope, contents and orientation of a scenario is based on a 'democratic' procedure where multiple and representative individual opinions and views are used to compose various (packages of) scenarios. Thus our scenario analysis is based on an aggregation of future images of various individuals.

Next, the scenario approach adopted here is explicitly *expert-based*. This means that the views of hundreds of transportation and environmental experts are used to compose new futures in which environmentally friendly transport systems play a critical role. The background of this approach is based on the fact that the field of sustainable transport is so complex and multifaceted, that a considerable knowledge and scientific insight is required to make solid statements on sustainable futures. Such experts may include transportation scientists, environmental scientists, planners, policy analysts, social psychologists and transport policy makers. It may be expected that such a broad and heterogeneous group of experts is able to make meaningful statements on sustainable transportation futures.

Finally, our scenario experiment is *menu-driven*, which means that experts are confronted withindividual features and driving forces which altogether make up a transportation system. By asking their views on these individual constituents of scenarios in terms of relative importance, significance, etc., it is possible to compose scenarios based on different compositions of the menus created by the experts at hand. This also allows for contrast analysis between different groups of experts (e.g. age, gender, disciplinary background, profession, etc.). In this way a broad spectrum of varying scenarios can be developed, while at the same time experts can be asked to assess the environmental, spatial or economic impacts of each of such scenarios.

An important point to be mentioned here is whether the above-described scenario design will ensure *consistent* scenarios. This is, of course, a difficult question. Consistency may have different meanings. It may refer to the fulfilment of a set of logical requirements in an underlying scenario model structure. If such a model does not exist (which is usually the case), consistency is not a meaningful concept from a strictly logical viewpoint, but refers rather to plausible, mutually compatible impact structures. In this context, it seems reasonable to adopt an expert-based approach, as experts may be assumed to be aware of underlying functional structures and mechanisms in a complex spatial transportation network. This approach is different from policy consistency; if there are multiple and conflicting objectives in transport policy making, unambiguous policy solutions cannot be found and have to be based on compromises which are never stable in a long-term perspective. But scenarios can highlight the consequences and the indigenous conflicts of such developments.

The above methodology will be used in subsequent chapters to describe the nature and results of our scenario analysis for sustainable transport systems.

7.5 CONCLUSIONS

The methodology of constructing scenarios may provide many new insights into possible developments, and may be very useful for investigating possible sustainable futures for the transport system. It is clear that in future transport systems many changes are necessary from the viewpoint of sustainable development, though as yet there is no clear picture of what an environmentally more friendly or sustainable transport system may look like. As a consequence, many different policy proposals and packages may be introduced in such scenarios (see also Thord, 1993). Thus, it appears that there are several types of scenarios, so that a choice should be made concerning which type to construct for a given purpose. This choice depends on the aim of the scenario, the budget, data availability, etc.

The purpose and advantage of scenarios is that they are a kind of structured brainstorming technique, which may widen the perceptions of researchers as well as policy makers regarding future possibilities and policy options and other related developments, and they may also give more insight into their impacts. In addition, they may increase the mental willingness to at least consider a broad spectrum of developments, since the resulting future is presented in an attractive - and undogmatic - way. In this way scenarios may be an interesting tool for strategic policy analysis, especially because policy makers often have too much biased and unstructured information. Scenarios may provide a clear and easily accessible - although not detailed - insight into policies and their impacts by constructing and evaluating several policy choices and packages. Also the sensitivity of the outcomes to changes in policies may be made clearer.

It may thus be concluded that the scenario methodology is a very interesting way of analysing the future of uncertain situations, and hence also for analysing the complex future of transport systems. In the next chapters we will therefore construct several scenarios, based on the methodology described above, focusing on sustainable transport systems of the next generation.

8 THE SPIDER MODEL FOR COMPOSING SCENARIOS

8.1 INTRODUCTION

Scenario analysis needs a frame of reference for judging the results or outcomes of alternative futures. In our scenario experiments we will use two contrasting views of the future in order to evaluate the outcomes of our expert-based scenarios. We will first develop a dedicated analytical tool, named the *'Spider model'*, to provide a web on which to build the reference scenarios and the expert-based scenarios in a systematic way.

Next, we will use the Spider model to focus our attention on the design and substance of the (extreme) reference scenarios for the transport system for the next generation (taking 2030 as the future reference year). These scenarios are assumed to be extreme future outlooks of society, its spatial organization, the resulting technical developments and related transport systems. They are partly descriptive, partly normative and mainly based on intuitive backcasting. Clearly, in this context also various other scenario experiments have been used for generating ideas on the future of transport (Biéber *et al.*, 1994; CPB, 1992b; Clerx and Verroen, 1992; ECMT, 1992; Grontmij, 1994; Masser *et al.*, 1992 and 1993; Schoemaker *et al.*, 1988; Svidén, 1990; Swahn *et al.*, 1994). Our extreme scenarios are meant to serve as a frame of reference for the series of expert-based scenarios which will be constructed in Chapter 9 and subsequent chapters. These may be expected to be less extreme and more of an intermediate nature than the reference scenarios.

In this chapter, we first discuss the methodology of the Spider model, and next the way transport scenarios are constructed by means of this approach. Finally, we describe our two reference scenarios, viz. a *market-based* and a *regulatory* one.

8.2 THE SPIDER METHODOLOGY

As discussed in previous chapters, there are many trends and driving forces which may impact on the future of transport. Such framing conditions may be found in several scientific fields and may relate in particular to spatial, institutional, economic and socio-psychological aspects. In the scenarios designed in our study, the resulting future transport system (including the transportation

technology) is supposed to be a response to forces and developments in these four fields, based on the analysis and information offered in previous chapters.
In this section we first discuss the methodology used for scenario construction - the Spider model - and then explain the way the scenarios are actually composed.

8.2.1 The Spider model

In Chapter 4 we discussed the various force fields influencing transportation technology and transportation systems with a view to assessing the sustainability implications of such developments. This can be summarized and depicted in a systematic way by means of Figure 8.1.

The most important future developments in these fields may be studied by using a qualitative type of multi-criteria approach, which can be visualized by means of a multidimensional Spider model (see Figure 8.2). Such a multi-criteria analysis is used here as a method to grasp, classify, analyse and evaluate different scenarios on the basis of explicitly formulated criteria (which are represented on each axis of the Spider).

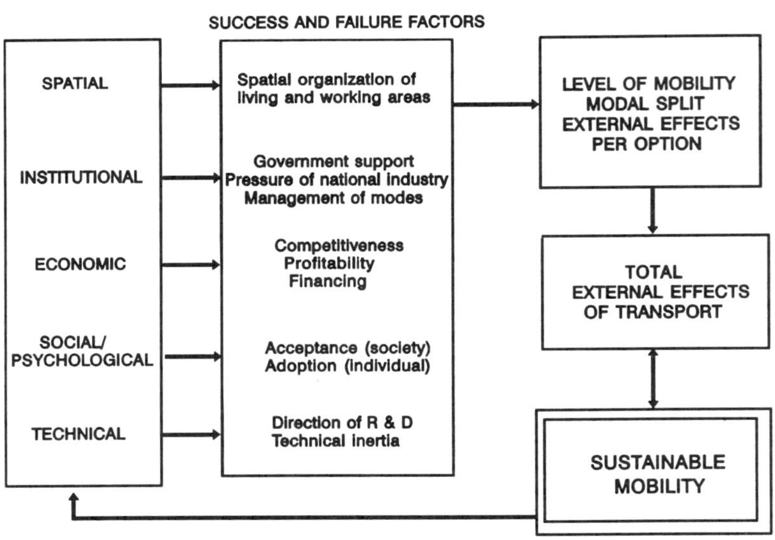

Figure 8.1. Success and failure factors of transport modes in various force fields

The advantage of this analysis is that the individual assessment criteria do not have to be measured in a single quantitative unit; they may be qualitative in nature (e.g. rank orders) (Nijkamp *et al.*, 1990; Voogd, 1983).

Each of the four relevant force fields distinguished in our study (i.e. spatial, institutional, economic and socio-psychological), which are seen as the most important factors influencing the future of transport and the transportation technologies used, are found in the Spider. As discussed in Part A, these factors are framing conditions (or driving forces), which are influencing, e.g. the mobility level, the modal split, the technical composition of the transport system, etc. The rank order of scores for the items on each of the eight axes is such that the inner points close to the origin are more associated with non-intervention (and, when necessary, market-based) strategies (e.g. market forces, liberal attitudes, etc.), whereas the outer points reflect the motives associated with policy (regulatory) interventions (e.g. land-use planning, control strategies, regulatory measures, etc.). We now discuss in more detail each quadrant and the relevant axes of the Spider diagram.

First, we will focus on the upper left corner, which contains the *spatial aspects* of transport systems. On Axis 1, possible developments in the future *European spatial organization* are mapped. At one extreme (the inner limit) of this axis we find the 'chains and zones model'. This is the end-result of a non-interventionist policy and of trends which lead to a diffuse spatial structure with many uncontrolled criss-cross transport links (see Section 3.5.2). At the other extreme (the outer limit), the concentration and specialization model is depicted. This tends to favour a concentration of activities and population in urban areas and in the economic core zone as a result of a dedicated spatial-economic policy aiming at coping with suburbanization and de-urbanization movements by imposing measures that guarantee more concentrated ways of living and working. Axis 2 in the spatial organization quadrant is mainly concerned with *urban patterns*; it describes the important dichotomy between spatial dispersion and concentration. Thus, on the two extreme ends of Axis 2 we have depicted the diffuse city (as a result of free market forces) and the compact city (as a result of explicit urban policy efforts).

In the *institutional part* (Axes 3 and 4 in the upper right quadrant) the control and management systems for sustainable mobility are represented. On Axis 3 the degree of *government intervention* in the transport market is depicted; the contrasting policies on this axis are regulation versus market-based measures. Next, the *management* of transport modes and infrastructure is found on Axis 4, where the two extreme ways in which this may be organized - as purely public or private provision - are presented. In this context, public transport means that the ownership as well as the operation of the transport companies are the sole responsibility of the government (or public agencies).

The third important driving force stems from the *economic field*, and is depicted on Axes 5 and 6. Axis 5 concerns the economic *feasibility* of transport, as the required profitability (at least cost coverage) of transport modes is

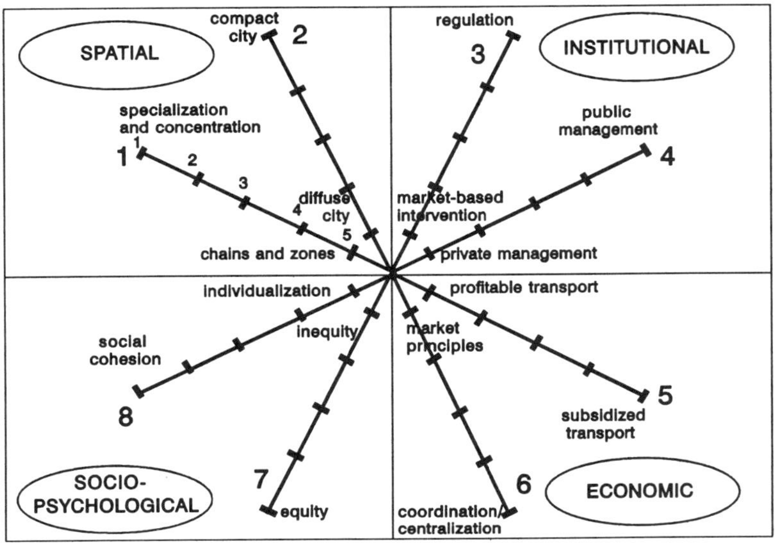

Figure 8.2. The Spider model for depicting the driving forces of future transport systems

an important factor for their chance to survive in the future. The main question in this case is whether the government wants to subsidize transport or whether transport modes should be operated on a commercial basis; in this way private financing of infrastructure may also be attractive. This justifies the extreme cases of profitable versus subsidized transport on Axis 5. The next extremes - to be found on Axis 6 - are the introduction of *market principles* in the economy - and hence also in the transport sector - versus coordination by the government (which may lead to some form of centralized government). This consideration is especially important for future economic growth, regional development, the construction of infrastructure towards peripheral regions, etc.

Finally, *socio-psychological factors* make up an important force field, which is depicted in the lower left quadrant. On Axis 7 the concept of *equity* is confronted with inequity. Non-intervention may favour inequity in society (for example, an unequal income distribution, unequal chances for individuals regarding, for instance, travelling or education), while on the other hand much public governance may emphasize equity measures (social security, access to transport services, discounts on travel costs for special groups, etc.). Finally, Axis 8 is concerned with *societal concern* and reflects individualization versus social cohesion. This distinction is important from the viewpoint of the level of freedom of individuals (versus social control), and includes related developments (e.g. demographic, educational), which may again have far-reaching consequences for transport.

All the above-mentioned fields (each represented by two axes) may exert important impacts on the structure of the transport market, the level of transport demand, the price level of transport and the modes used, and may hence be regarded as the main force fields for the development of transport systems technology. As mentioned before, technological advances in the transport sector are mainly regarded here as responses to these driving forces ('market pull' strategies), which serve to cope with the challenges imposed by the need for sustainable transport systems ('technology push'). Hence, they are not represented on separate axes of the Spider diagram.

8.2.2 The construction of scenarios

A transport system can now be represented and assessed by a combination of eight points (or qualitative rank order scores) on the successive axes of the Spider model. This is a meaningful visualization of the main characteristics and driving forces of such a system, as a confrontation of different 'Spiders' (concerned with different driving forces or different actors) will immediately pinpoint the most important underlying factors. It should be recognized that the size of the area formed by linking the eight points on all axes has no meaning, because:

- the information on the axes has only a qualitative (and not a cardinal) meaning;
- the size of the resulting area is also dependent on the order in which the axes are positioned in the Spider.

It should once more be emphasized that technological developments are not included as autonomous driving forces in the Spider model. Clearly, it would certainly have been possible to include technology in the Spider web (e.g. by making a distinction between entrepreneurial types of technological changes and government-instigated types). However, in the present study, technological advances in the transport sector are regarded as endogenous developments (i.e. responses) to the underlying four major driving forces. Thus, in our approach transportation technology is not 'manna from heaven', but a result of systemic forces. Consequently, in the present study a typology of various technological options in the transport sector is made (see Part B), and then each of these options can be evaluated by means of the elements from the Spider model.

It should also be explicitly mentioned that the two extreme points on each axis have only a qualitative meaning; they do not represent numerical information, but only a rank order (in terms of more or less). This is also important for scenario design, as the expert-based statements on the successive force fields are more concerned with the likelihood of occurrence of transport systems options and underlying forces than with precise assessments of all consequences of such options (see Chapters 9 and 10). It is possible - or even likely - that in practice the expected future developments will be less extreme than the inner or outer limits. In that case a shrinkage along axes may take place in order to describe such actual developments. Thus, the central point (point 3) on each axis indicates an intermediate (neutral) development, while points 2 and 4 represent developments which are closer to one of the extremes.

The two reference scenarios in our investigation of future transport systems are constructed in such a way that they form the inner and outer circle of the Spider. They mainly serve as a frame of reference for assessing and evaluating intermediate scenarios. The expert scenarios will probably combine points on the axes which are positioned more at the middle of the axes. Clearly, there are numerous such scenarios, so that a wide variety can in principle be constructed depending on the views of particular groups of experts. This will be further discussed in Chapter 10. First, however, we will describe the extreme reference scenarios in greater detail.

8.3 THE MARKET-ORIENTED SCENARIO

In the market-oriented scenario it is assumed that the spatial organization will develop according to the inner circle of the Spider model (see Figure 8.3).

The main assumptions about the spatial organization are that there will be a development according to the chains and zones model and towards diffuse cities. In the market-oriented scenario it is also assumed that spatial patterns develop with a low level of concentration (density), while government policy will be mainly market-oriented and focused on reducing its influence on society and the economy.

As a result government intervention in the economy will be as small as possible, while the necessary (especially environmental) measures will be market-oriented. In subsections 8.3.1-8.3.5) we illustrate the substance and functioning of the transport market in the market-oriented scenario by means of an imaginative description of the future in the year 2030. This scientific story or picture is consistent with a market view on transport, but, of course, does not claim to be a 100 per cent accurate representation of the future. Each of the quadrants and axes of the Spider model are now dealt with.

8.3.1 Spatial factors

The European spatial organization

The spatial organization is the starting-point of the analysis for the market-oriented scenario for the year 2030. The suburbanization trend will continue, so that a further shift of population and activities out of the cities

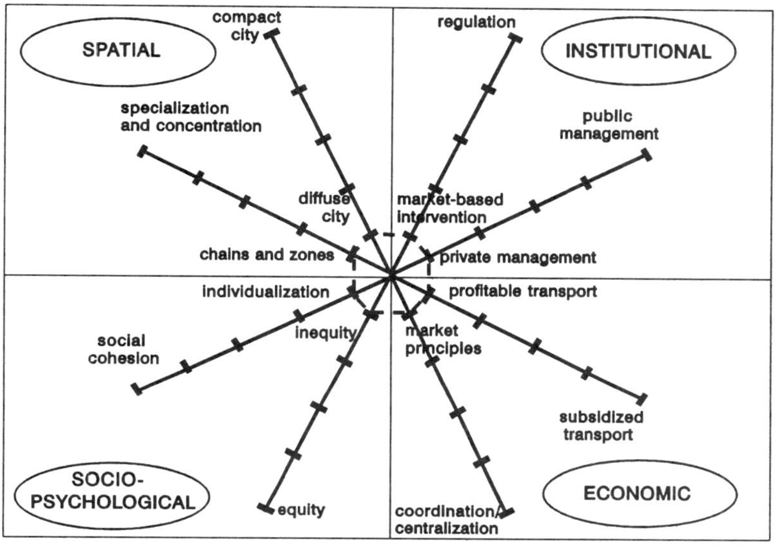

Figure 8.3. The Spider model: the market-oriented scenario

will take place. In the core regions a more diffuse spatial organization will therefore develop, but there will also be a shift out of the core regions to outer regions (including rural and green areas). These regions appear to be very attractive, because of their desirable living conditions, the availability of relatively cheap premises and low levels of congestion. Because of the high price of mobility, however, only those with higher incomes can afford to move out to these regions, whereas the lower income groups are forced to stay mainly in the cities. Governments will not interfere in this development, because it is argued that each individual is best-placed to freely choose his or her place of living.

Accordingly, the working areas will also be positioned in a more diffuse way, because the trend towards footloose companies will continue, which makes this development possible. It will be attractive for new companies, which need a highly educated working force, to move towards the living areas, in order to be located close to their labour force. In particular, dynamic new service companies may move towards these regions, while the manufacturing sector - which is less dynamic and needs a less educated labour force - will stay behind in or near the urban areas. Those areas which can be reached by high- quality infrastructure will be the most popular for living and working.

As a result, rural regions appear to be very competitive compared to urban areas, despite the higher price of mobility. The support for large cities will largely be abandoned, and government policy will not alter this trend.

High-quality road infrastructure - perhaps even underground - may be constructed towards these regions. As a result the population will grow there, while it will stabilize or decline in the cities and the core regions. In this way a development according to the chains and zones model will occur, with many criss-cross transport links and a relatively low level of demand on each transport link.

The spatial organization at the urban level
Every region will have one or more cities, which are the main centres of population and economic activity. A region will be self-supporting for many services and production activities, because of the high transport costs. Because of the decentralization, the regions will also become politically much more independent. As mentioned above, the relative economic position of the cities will decline, because more highly educated people and many dynamic firms will move out of the cities. Although the size of these cities may largely differ, one cannot speak of a classic hierarchical spatial organization.

The spatial organization of cities will develop in a diffuse way, while the population of most cities will not grow. Because of the individualization, the size of households will decrease and therefore more houses are needed. The people in higher income groups who still prefer to live in cities, will largely move towards 'green' suburbs. Therefore, the cities will spread out further, while the population density of the cities will decline. Also shopping malls and other services will shift to the suburbs, while the city centres will partly lose their central function, because of the high mobility costs and parking fees. The lower income groups may stay behind in the old living quarters, where cheap but low-quality housing is available. In these quarters little investment will take place, because the government will have largely abandoned its housing policy. Working areas will also be spread out over the cities, because there will be no active government policy to stop this trend.

8.3.2 Institutional factors

Intervention by the government
The market-oriented scenario takes for granted that the devolution trend will continue in the institutional field. Organizations like the UN, the OECD or the World Bank will lose part of their strategic position, while national states will gain influence. There may be one major exception to this trend, however. Because more and more environmental problems will manifest themselves at the international level (e.g. the greenhouse effect, depletion of the ozone layer and acid rain), it is not possible for individual countries to solve these problems. Therefore, a global consensus will be reached in which a strict environmental policy is agreed and imposed all over the world. 'Free rider' behaviour of individual countries - by avoiding strict environmental measures in order to attract

the economic activities of other countries - will not be possible because of joint sanctions exercised by the other countries. For example, a worldwide CO_2 tax will be introduced by the beginning of the next century, as well as taxes on other emissions, in combination with a system of tradeable emission rights. This will significantly increase the price of fossil fuels, which will certainly reduce their use. As a result, the CO_2 emissions and other emissions will diminish drastically, while public revenues will be mainly based on environmental taxes.

The European countries will also participate in this international agreement. In other fields, however, a reversal in European integration is likely to occur. The EU will fail to integrate all European countries; such a disintegration means that regional policies as well as subsidies to the agricultural sector will generally be abandoned. Most emphasis will be put on strong national economic development and on the competiveness of countries and regions. This will also reinforce the trend of governments to 'step back'. The regions in Europe will become more independent and self-supporting, because at the national level many policy fields will also be decentralized. There will be a strong decline in active housing, spatial and transport policy. This is all left to the market although external costs will be internalized.

Organization of the transport market

As a result of the above-mentioned policies, governments will mainly withdraw from the transport market and the public provision of infrastructure. The responsibility for European railway companies, for example, will be left to private parties. Therefore, they will be operated on a commercial basis, which may drastically increase the fares.

Infrastructure may be owned, financed and operated by private parties, too. Many user charges will be introduced to pay for the infrastructure; for example, electronic road pricing may be introduced all over Europe. The management and operation of rail infrastructure will also be separated. Single operators will be replaced by many operators competing in the same market. User charges will make investments profitable everywhere; if demand is not sufficient to cover the costs, the infrastructure will simply be closed down.

Since much new infrastructure will be constructed underground, because of the lower negative externalities and therefore lower environmental taxes, governmental influence in the planning of infrastructure will also substantially decline.

8.3.3 Economic factors

Organization and development of the European economy

In the market-driven scenario, the European economy will show a relatively favourable development, since much attention will be paid to the introduction of incentives and other market principles in the economy. This will make the European economy more competitive compared to other trade blocks and stimulate economic growth. Because of the high transport costs, however, trade growth will be lower than current growth rates. As a result of this and the European spatial organization, regions and cities will become more self-supporting. Another consequence of these policies will be that the income distribution will become much more unequal. Because of the high economic growth, however, the average income will tend to increase; and therefore, many people can - despite the high mobility prices - afford to live outside the cities in rural areas.

Regional economic growth will also spread to other regions because of the new European spatial development. As a result of the shift of well-educated people and dynamic firms towards the peripheral and rural regions, the economies of these regions will develop in a favourable way. The core regions, on the other hand, will face a relative economic decline, because of the shift of wealthier people and companies out of these regions. Cities will decline, too, because high income groups and many firms in the tertiary sector will move out of the cities. No central government support will be given to the cities, however, because of the general trend for central governments to step back and favour fiscal decentralization.

Profitability of transport modes

Public policy will be aiming at internalizing environmental costs by increasing the costs of fossil fuels by means of CO_2 taxes and environmental taxes on other emissions and external effects like noise and stench annoyance, and visual intrusion. These taxes will be introduced stepwise at the beginning of the next century and will appear to be so high that the development of alternative fuels and new modes becomes profitable. Fossil fuels will only be used if really necessary. Because of technical innovations, the use of fossil fuels will diminish, which will significantly reduce the emissions of CO_2 and other greenhouse gases.

The variable costs of mobility will also increase by privatizing infrastructure. Road-pricing and cordon-pricing systems for road infrastructure will be introduced in all European countries; the revenues of these charges will be used for construction, maintenance and profits by the infrastructure owners and operators. An additional charge for external effects like noise and stench annoyance, and visual intrusion will be levied on these prices. Rail infrastructure will also be privatized, and therefore all rail links will be open for all companies.

All public transport will be privatized too. All links will in principle be profitable. Regulations will diminish and only basic safety standards will have to be met. Since uneconomic links will be closed, all transport in Europe will become profitable in this way.

8.3.4 Socio-psychological factors

Individualization

There is, in the market-oriented scenario, widespread agreement in society that the environment should be protected, and therefore strong measures are accepted in society. This will be the case all over the world. Therefore, such measures at even a global level will be possible, despite the trend towards the disintegration of the world community. The individualization trend will become more pervasive at the same time. From past experiences it is clear that despite the general acceptance in society, individuals will try to get round environmental problems and related financial burdens by means of evasion (free-rider behaviour). As a consequence, only market measures will be introduced; evasion is much more difficult then. The market measures will internalize external costs, for example, in the price of transport.

It will be argued that every individual should be free to choose his or her living area, and therefore the richer social classes will move out of the cities, while in the older city centres the poorer people will stay behind. A similar situation will hold for transport, where everyone is free to choose his or her own way of transport. If the external costs become too high, taxes are raised, after which the market will solve the allocation problem. In this way the travel possibilities of individuals will be restricted by their income level.

Equity in society

Social security systems will largely be privatized and minimized; the same holds for all kinds of government subsidies and support. Therefore, a strong polarization of society will take place, as the income distribution will become much more unequal. As a result, the possibilities for mobility and housing will become much more unequal too.

Society will be subdivided into three classes: the first class is that of rich people, which includes about 15 to 25 per cent of the population. They will own a private car and are not restricted by the high costs of mobility, for example, of air traffic. Therefore, they will be able to live where they want to, which is often in green areas with a high landscape value.

The second class is the middle class, which may include 55 to 65 per cent of the population. They can also afford a private car, although travelling will be relatively expensive. They will mostly live in suburbs and in rural regions. Teleworking and telemeetings will become the working habit for these people for a few days a week, which will reduce their mobility needs. Also working days will be made longer, though people will have more holidays too.

The third class is the less privileged class of mostly or low-income unemployed people (about 15 to 25 per cent of the population), who cannot - or can hardly - afford a private car and will be dependent on collective transport modes, car rental services, etc. Because of their low income level, they can only afford to live in the old and low quality city quarters.

8.3.5 The transport system in the market-oriented scenario

Long-distance transport

As a result of the diffuse spatial organization in our market-driven scenario many low demand criss-cross transport relations will emerge. Individuals will be free in their choice of transport mode, although mobility will become much more expensive, and hence transport demand will be restricted by the income of individuals. Taxes on CO_2 emissions (and other externalities) on modes and infrastructure will be introduced stepwise from the beginning of the next century, which will make fossil fuels relatively unattractive. Alternative fuels will therefore be introduced on a large scale. Infrastructure will be privatized and planned, constructed and operated by the private sector. Collective modes will be privatized too and those links which are not profitable, will largely be closed.

Because of the many low demand links in this spatial configuration, individual modes will be the most efficient way of transport in most cases. The middle and upper classes will own a private car in general, which is powered by non-fossil fuels. These private cars will be guided by telematics systems, which will provide the most efficient and cheapest routes to their users. Since the costs of mobility will be quite high, the middle class will reduce its transport demand by teleworking and related services. On high density links physical route guidance of cars will become common practice in one generation from now, because in this way the infrastructure will be used in the most efficient and thus the cheapest way. This will increase the capacity of the infrastructure significantly, which will induce a further rise in mobility.

Surface infrastructure will be heavily taxed because of the negative externalities involved. Therefore, it will become attractive to construct subterranean infrastructure for long distances, and moreover this technology will become much cheaper. The same also holds for existing highways, as their use will become relatively expensive because of the higher external costs. Therefore, there will also be a trend to construct these routes underground as far as possible (e.g. by tunnelling).

At the surface mainly only local or regional traffic will be found, which will cause less annoyance. Since it is not likely that the rail infrastructure will be constructed as a subterranean network, on several links the existing lines will be closed down. Bus services - which use the already-existing road infrastructure - will then replace these services. Only high density links, collective modes like HST and Maglev high-speed trains will be in operation. This will be the case only on links between the larger cities, where there is a high level of demand. On these links the use of these modes will be attractive because of the lower costs, the high speed and the high service level. There will also be a widespread possibility to transport cars in these trains.

It is likely that no - or few - trains will serve the smaller cities, because of the closing down of several links. A bus system will be set up which offers services, mainly used by people who do not own a car. These buses will only offer a low quality service, because of the low level of demand.

On very long distances, where no good HST or Maglev connection is available, LH_2 planes will offer an expensive service, since this transport mode will be heavily taxed because of the high external costs. As a result, these planes will be mainly used by the higher income groups.

In the first epoch of the next century mobility growth may even become negative, because of the new environmental taxes. Later on, the transport system will adjust to the new price structure, and new technologies will be introduced, which will become cheaper too. As a result, prices will be lowered, which together with continued economic growth and a more diffuse spatial structure, will cause new growth in mobility. New infrastructure, however, will mainly be constructed underground, because of the low externalities; so there will in fact be no need to reduce mobility levels any further.

The modal split on long-distance routes will change further in favour of the private car. The modal share of the latter mode will be high (probably about 90 per cent), while almost all cars will run on new fuels. Collective modes will therefore have only a marginal modal share.

Urban level

The cities will develop towards a diffuse spatial structure, while the population will not significantly increase. Therefore, many low-density, criss-cross transport relations will emerge. Public transport in the cities will be privatized. Because of the low level of fares, however, many links will appear to be unprofitable, so that many collective transport links will be axed. For people who do not own a private car, an extensive bus system will be in operation, which may offer a low-quality service to most destinations in the cities.

Private cars will, therefore, be the main mode of transport, while in various cities on shorter distances the bicycle will also be an important mode. Only along some high-density links will subterranean people movers offer a more frequent service. As with long-distance transport, the main corridors will gradually be turned into subterranean links, while the cars will be guided physically in order to increase the capacity of the infrastructure. A user charge will be demanded on all infrastructure by means of cordon and road pricing; this will, of course, increase the price of urban mobility. Surface parking will become very expensive, too, and therefore many subterranean parking garages which are privately operated will be constructed. On the surface only low density traffic flows will be found, mainly in the most important living quarters of cities.

Because of the high price of mobility, the stabilizing population and a shift of economic activities towards extra-urban places, urban mobility will decrease in the next century, or at least stabilize around the year 2010. As a result, little new infrastructure will be constructed. The modal share of the private car in cities will be relatively high (some 60 per cent), while the bicycle and the bus will have a reasonable share.

Thus the market-oriented scenario seems to be a straightforward extension of free mobility behaviour, although constrained by the price mechanism to ensure environmental sustainability.

8.4 THE REGULATORY SCENARIO

The starting-point of the regulatory scenario is the spatial specialization and concentration model at the European spatial scale level and the development of compact cities at the urban scale level. In this scenario it is assumed that the government will mainly use regulatory measures to influence the transport market, because of a political emphasis on achieving equity in society. This scenario forms the outer circle of the Spider model and is thus the complete opposite of the market-oriented scenario (see Figure 8.4).

8.4.1 Spatial factors

The European spatial organization

In the regulatory scenario it is taken for granted that by the end of the 1990s the suburbanization trend will reverse. This will be one of the consequences of a spatial policy, which will aim at concentrating activities in the cities and specific areas in order to achieve a compact use of space. As a result, a movement of activities and population back to the cities will start to take place.

Leading economic activities will be found in several metropolises, mainly as a result of the fact that business firms will become less footloose because of the dependence on the public sector and on collective modes. Most of these metropolises will house a large population (ranging from 10 to 20 million) and will be the centres of economic activities. In these cities, besides regional services, several head offices of international companies, financial markets, large industries, junctions of HST/Maglev-connections, and a mainport (large harbour and/or airport) will be found. In Western Europe these metropolises are notably London, Paris, Frankfurt, Milan, Barcelona and the Randstad Holland. This trend will be reinforced by the fact that goods transport will become expensive and inefficient, since the use of lorries will be severely restricted. So the spatial organization will develop towards the specialization and concentration model, with metropolises at the top of this hierarchical system.

One level down in the hierarchy, several 'Europoles' will develop. These will be smaller than the metropolises, but will still be large cities (1 to 5 million inhabitants). They will be oriented towards the metropolises for many services, but they will house many main activities themselves, for example, an HST connection, a regional harbour and/or an airport. They play an important role in the European space as intermediate nodes. At a lower level in the spatial organization, the smaller cities will be found; these are centres in which regional activities will be concentrated.

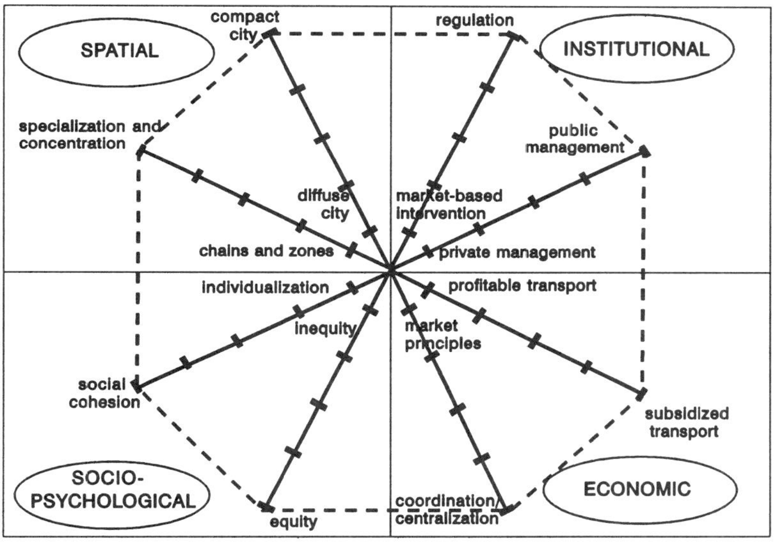

Figure 8.4. The Spider model: the regulatory scenario

This spatial development will have negative consequences for rural areas as well as for peripheral regions, as the population in these areas will decrease considerably, since public policy will force activities and population to shift to the cities and other dedicated nodes. Agriculture, for example, will now mainly be found around the cities, because of the inefficient transport system. In many (former) rural regions land will no longer be needed for agriculture, because of the rise in productivity, together with a stabilization of the population. This trend will be reinforced by policies of the government to create nature parks from several rural areas.

The urban spatial organization
All cities will develop towards the compact city model, with a concentration of

economic activities and population in the city centres and some subcentres as a result of the above-mentioned public policy. This will become manifest in every city type (metropolis, Europolis, etc.). A lack of space will occur in the city centres and around main nodes of the collective modes, however, which will make space very expensive there. This will stimulate the construction of both high-rise and subterranean buildings. The main reason for this development is that it will be efficient and attractive to live or be located near high-quality nodes of collective transport modes, because in other areas travelling will become too expensive and time-consuming.

In old living quarters of the large cities we will find mainly those people who cannot afford the housing prices in the centre or subcentres. In these quarters housing and public transport will be of a relatively low quality, although every place in town can be reached by public transport.

8.4.2 Institutional factors

Intervention by the government
The devolution trend will reverse, because it appears that this causes too much inequity in society. Much emphasis will therefore be put on regulatory measures, because then less distinction is made between different socio-economic groups. The European Union will develop towards a centralist organization with an abundance of regulatory power. A strict environmental policy, for example, will be introduced, which will be obligatory for every Member State. To reduce the external effects caused by transport, many regulations will be introduced, e.g. compulsory targets for CO_2 and other emissions. It will also be forbidden to build living and working areas in rural and peripheral areas, in order to reduce the mobility level and to concentrate traffic flows. Furthermore, many protectionist measures will be introduced to safeguard the European economy and to make it possible to introduce the regulatory measures.

Most attention will be paid to equity: social security systems and government interventions will be expanded in almost every sector of the economy, and peripheral and rural regions will to a large extent be subsidized. All these measures will significantly increase the influence of governments on society and the economy, while private sector involvement and market incentives will be reduced.

Organization of the transport market
The government will regulate the whole transport market because of the high external effects of transport. For example, unlimited private car use will generally be forbidden for equity reasons, as well as because of the high external costs of cars. Therefore, collective modes will become dominant and the infrastructure will drastically be expanded.

The railway companies will all be publicly-owned, and the trend towards decentralization and privatization of the companies will reverse. They will grow into bureaucratic organizations, operating at substantial losses. As a result, transport will not be organized in an efficient way but, since the companies will face no competition and the losses are covered by the government, this is accepted because of the political emphasis on equality. The government will also be responsible for the infrastructure; this holds for planning, financing as well as maintenance.

8.4.3 Economic factors

Development and organization of the European economy

The development of the European economy will lag behind developments in the United States and the Pacific region, because most attention will be paid to equity in society and to regulatory measures. Market principles will mainly be abolished to achieve these goals; this will reduce economic efficiency and, as a result, economic growth. To protect employment in Europe and to introduce environmental measures, high tariffs and other protectionist measures will be introduced, and also taxes in general will be increased. This will again have negative consequences for economic growth.

The EU will not be able to achieve an evenly spread regional development, nor will it be able to ensure a rapid economic development in peripheral regions, since in that case the desired concentration of activities cannot be achieved. To keep these regions economically healthy, subsidies will be given to the population at large. This policy will cause a regional development in which the core regions relatively improve their economic position, while peripheral and rural regions face a decline.

Profitability of transport modes

In order to save space, to decrease external environmental effects and to stimulate the use of collective modes, the use of cars run on alternative fuels will be regulated and stimulated. This may be done by allowing everyone to drive only a limited amount of kilometres. For reasons of equity, the prices of transport by collective modes will be kept low, and because of the many links to be covered and the low prices, they will operate at big losses, which will be covered by the governments.

The same will hold for urban transport. The use of private cars will be strongly discouraged, while public transport connections will be constructed as subterranean infrastructure on a large scale. Almost every destination can be reached by public transport because of the high density of the network. Therefore demand on many links will be low, which will cause huge losses for public operators.

8.4.4 Socio-psychological factors

Individualization

The lower economic growth, together with the growing emphasis on equity, will to a great extent reverse the individualization trend. It will be clear that environmental measures are necessary, and it is widely accepted that this will be done in a regulatory way. As a result social cohesion will increase, and evasion may therefore cause little problems.

The spatial organization will force people to live in relatively small dwellings because of the lack of space. This is viable, as the average number of people per household will also increase.

Equity in society

Equity - besides environment - will be a main objective of government policy. For example, social security systems will be expanded, while government subsidies and support will increase everywhere. Mobility, for example, will be kept cheap by providing subsidies. As a result inequalities in society, like income distribution, will decrease. The same will hold for regional development because of the large subsidies.

Three income groups can be distinguished. The wealthy people will account for a small share (probably some 5 per cent) of the population, who will mostly live outside the cities in high-quality housing areas. When necessary, these people will also drive cars using alternative fuels. They may travel an unlimited number of kilometres, when they need the car for their work or business and for safety or other reasons. This privileged lifestyle will be accepted despite the emphasis on equity in society.

The largest share of the population (about 90 per cent) of the population will belong to the middle-income group. These people will mainly live in the centres and subcentres of cities. They will mainly use collective modes for transport, which they can easily afford because of the low fares. They will mostly work in areas which can easily be reached by collective modes; the same will hold for shopping malls, etc. Teleworking and telemeetings will not grow to a large extent, since transport fares will be kept low, so tele-activities will not be very profitable. Recreation areas will mainly be located in the vicinity of living areas, around cities within easy reach of collective modes. Leisure traffic will therefore decrease. To reach nature reserves, however, car use will sometimes be necessary.

The last group (some 5 per cent of the population) will be found in the less privileged, low-income group, such as structurally unemployed people, guest workers, etc., although government policies will be developed to integrate them. They will live outside city centres (and other subcentres) in old low-quality living quarters.

8.4.5 The transport system in the regulatory scenario

Long-distance transport

At the European level a radial transport system will emerge in the regulating scenario, with high density transport flows between the metropolises and from the metropolises to the Europoles. Car use will largely be regulated, because the ownership of cars is strictly discouraged, while car rental will be only allowed for a certain maximum level of kilometres per person. These measures will be introduced step by step from the beginning of the next century. In the same period infrastructure for collective modes will be constructed everywhere, the use of which will also be stimulated by the spatial organization of the society. The infrastructure will often be constructed underground at sites formerly occupied by road infrastructure, now out of use. Therefore, transport demand for collective modes will largely increase, while private car use will decrease.

Between the metropolises (the transport links with the highest demand) several subterranean high-speed shuttles will be constructed, which offer a high-quality service. Between the other metropolises and between metropolises and Europoles, high-quality HST and Maglev connections will offer a frequent service. This system will be the backbone of the transport system. From the metropolises and the Europoles fast train connections to the smaller cities will also be available. These are slower than an HST, however, since there will be more stops. Their capacity will be much larger than trains nowadays.

To connect the smaller cities and other living and working areas, a system of improved conventional trains will be functioning, while for regional transport slower trains, subways and light rail connections will be in operation. Europe will be almost completely covered by a public transport network in this way.

As a result, a collective transport system will be constructed, which offers a high frequency and a good service at low prices, so that almost everyone can afford to travel. Although these collective systems favour a concentrated spatial organization, the system will still be very unprofitable. This will be caused by the low demand on many connections, together with the low prices, the high service level (high frequency, etc.), the inefficient organization of the public transport companies and the high density of the network. The operation will in general be subsidized by the government, while the system will be of lower quality than the 1990s private car system.

For people who live in places at a greater distance from nodes of public transport, travelling will become very difficult and expensive. This will especially be the case in less densely populated regions, a situation which will make it unattractive to live there.

On connections where efficient shuttle, HST and Maglev connections are available, air traffic will be strictly discouraged and regulated. On longer distances especially, LH_2 planes will offer a service, but it will be relatively expensive.

In the next century mobility growth with respect to long distances will be negative because of the regulations restricting private car use, the low economic growth and the stabilizing population. If the collective modes offer more connections, however, the mobility will rise again, but at a lower growth rate than in the 1980s and 1990s. In the first part of the next century, infrastructure for collective modes will be constructed on a large scale, but after this period such construction will decrease, although new links will still be built. In one generation from now, mobility growth will stop, because of the stabilizing population and the continuing spatial concentration.

The modal split will largely change in favour of collective modes. It is likely that about 80 per cent of transport demand on long distances will be covered by collective modes. The remaining 20 per cent will be covered by cars run on new fuels, of which a significant share will be rented and the remainder privately owned cars.

Urban level

In the cities a radial transport system will be developed, with the congested city centre as a central point where all high density transport flows come together. The main corridors will also connect the subcentres of activities, so several high density transport links will be found in city centres. The central transport nodes will be located at HST/Maglev stations in the bigger cities and at many train stations positioned in subcentres. Private car use in the cities will largely be abolished, because of the above-mentioned regulations. Therefore, almost all transport links will be covered by collective modes for longer distances, while walking and cycling will mainly be used for shorter distances.

An extensive network of public transport will be constructed. People movers will offer a high frequency service from the city centres to the subcentres. This network will be rather diffuse and forms the backbone of the urban transport system. At the nodes of this system lower quality modes like subways, zero-emission buses and trams will be in operation. At these nodes bicycles may also be rented easily, which will be used for destinations which are not easily reached by public transport. In most cases, there will be separate bicycle lanes. Moreover, the poorest quarters will be covered by a collective transport system (often of a relatively low quality), the use of which will be stimulated by discounts for people living on a minimum income.

Road infrastructure will mainly be used by buses and for walking and cycling, while private cars will only use a limited part of the infrastructure. The low fares (for equity reasons) will make the public transport system very unprofitable. The government will therefore subsidize transport to a great extent. Private cars will only be used - as an exception - by higher income people or will be rented for certain journeys. Rented vehicles will be used, for example, when heavy goods have to be transported, or to reach recreational areas.

Because of the low price of mobility the use of these collective modes will be restricted by limitations (in travel time, destination, etc.) instead of by the price level. Because of the high frequency and the low prices the system will not be profitable, although the spatial organization will be favourable from a sustainability perspective. The government will subsidize the system to a considerable extent, which will be accepted because of the low external effects and for reasons of equity.

The modal split of the private car (mainly electric cars) will dramatically diminish to a level as low as 10 per cent, while slow traffic and collective modes will account for the remaining 90 per cent. The mobility in cities will decrease in the next century because of the measures introduced. When many links of collective modes have been constructed and finalized, mobility will rise again, although the mobility level will remain lower than before, because the public transport system will offer a lower quality service than the private car. Later on in the next century the total mobility level will rise again, because of the growing population.

8.5 COMPARISON OF THE REFERENCE SCENARIOS

The two contrasting scenarios elaborated above each describe a different extreme development of the future, since the assumptions about the spatial organization and the emphasis on government policies are at odds with one another. Both scenarios result in a completely different transport system, the one dominated by individual modes and the other by collective transportation. In both scenarios the CO_2 emissions may be reduced substantially. In the market scenario most population and activities are found outside the cities, while in the regulatory scenario these are more concentrated in the compact cities, and therefore also the mobility patterns differ largely.

The advantages of the market scenario may be that a sustainable transport system develops without large-scale intervention of the government. The spatial organization and the transport system develop rather naturally because of market forces. This also has positive consequences for economic growth, and the economic development of regions is quite equal. Another advantage is that there is freedom of choice for individuals. Transport is operated in an efficient way, since the influence of the government is minimized.

Disadvantages, however, are found in the field of social equity. There is quite a large number of people who have relatively more negative living conditions, while they also have fewer possibilities for satisfying their transport needs. In this way transport may become a luxury good. In the regulatory scenario the advantages and disadvantages are largely the mirror image of the market-oriented scenario.

By using these extreme - though not necessarily realistic - reference scenarios, it is made clear that there may be different societal trends and public policies, which may both lead to a sustainable transport system. The degree to which each of these extreme scenarios would contribute to a reduction in CO_2 and other greenhouse gases, needs to be assessed before final conclusions can be drawn. This will be the subject of Chapter 12. Another point concerns the degree of acceptability of such contrasting scenarios.

In practice, it is questionable whether the extreme measures assumed in these scenarios could be accepted and introduced. In reality, less extreme scenarios are far more likely, as is borne out by the expert-based scenarios presented in Chapters 9 and 10.

9 EXPERT VIEWS ON THE FUTURE TRANSPORT SYSTEM

9.1 INTRODUCTION

A scenario experiment is not just a description of a dream. To have any credibility it has to be based on a combination of real-world insights/information and scientific imagination. In this way our Spider scenarios have been designed, in order to make them also suitable for testing transportation policy measures. In this chapter we present the views on future transport systems expressed by a scientific experts group. The main thrust is the construction of the foundation of a transport system from a blend of transportation experts views.

As discussed before (Section 3.9), we have used a representative sample of 271 Dutch transport experts to create future images and to compose scenarios on sustainable transport development in the next century. Therefore, after the construction and mapping out of the reference scenarios in the previous chapter, we will now analyse how experts view the future of transport systems. These views were obtained from an extensive postal survey questionnaire, containing a wide range of detailed questions, while the experts' own opinions could be added as new items to be considered. In this chapter the results from this survey will be discussed. It should be noted that in Section 3.9 various issues related to current trends and their consequences have already been dealt with. We will explore mainly behavioural and policy issues from the questionnaire as well as expectations and desires concerning the transport system in the year 2030.

The questions about the transport market in 2030 have been subdivided into three subsystems, which will be dealt with separately:

- the urban transport system (action radius of less than 20 km);
- the regional transport system (action radius ranging from 20 to 50 km);
- the long-distance transport system (action radius of more than 50 km).

After the analysis and presentation of the results from expert opinions in this chapter, in Chapter 10 we will construct two distinct scenarios - an expected and a desired one - using the results discussed in the present chapter.

9.2 EXPECTED AND DESIRED BEHAVIOURAL CHOICES AND POLICIES

9.2.1 Opinions about behavioural choices

Expected behaviour

As mentioned above, the opinions of the experts could mainly be expressed by using qualitative scores on each of the items under investigation. In this way, various frequencies could be derived on all revealed statements of the experts. When we investigate the expected options (see Figure 9.1), it appears that 'buying cleaner and more economical cars' is given by far the highest score. This is not surprising, since this option does not require a drastic change in mobility behaviour, and the same trend is also found in the past decades. The other options interfere more with the functioning of the present transport system and the mobility behaviour of individuals.

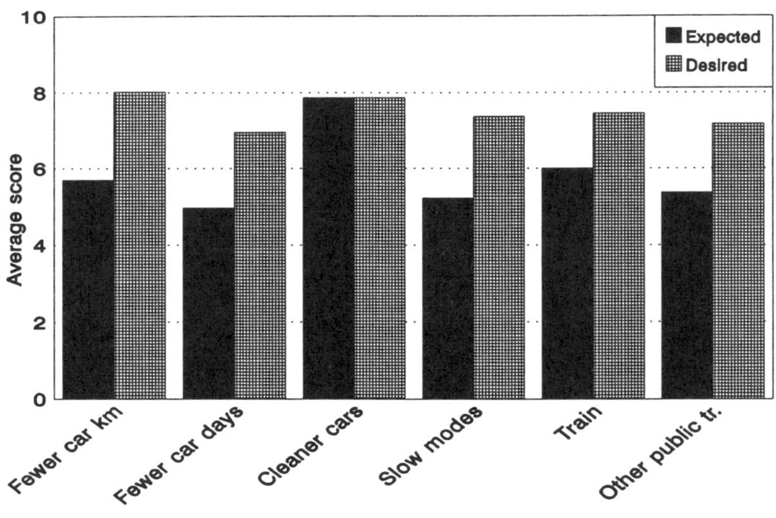

Figure 9.1. Expected and desired behavioural options

Notes: 1) Scores 1-10; 1 is lowest, 10 is highest

2) n = 246-259 (not all 271 respondents answered all questions)

In the second place - but with a much lower score - we find an expected increase in the use of trains. The other scores are very low (below 6), so that these options may not be expected to be freely accepted on a large scale. The third highest score is given to a reduction in the number of car kilometres. This was to be expected, since in several government campaigns and policy debates it has often been argued that it is not the ownership but the use of cars that should be reduced. Next in the rank order list comes the use of other public modes, which is lower than the use of trains. Finally, the lowest score is given to the option of using the car on fewer days. From this it may be concluded that the received wisdom that 'when one has a car, one uses it' may also be expected to hold in the future.

Additional options also mentioned a few times are: better transport management by public transport companies to provide collective transport (two times), using telematics (three times, although this option maybe partly incorporated in driving fewer kilometres) and a different spatial arrangement of society (for example, living closer to the work place, 10 times). The average score of the last option is low, however, viz. only 3.80.

Desired behaviour

When the desired behaviour is analysed, it appears in the first place that each option - except buying cleaner engines - gets a higher average score than the expected behaviour. Thus, for all options the expected behavioural change is less far-reaching than the desired one; and therefore, the foreseen transaction costs of the desired behavioural options are in fact much higher. It is also striking that the range of the scores is much smaller.

Reducing the amount of car kilometres now gets the highest score, which may be an indication of the fact that too many unnecessary kilometres are driven by car users. So in theory the above-mentioned government policy - aimed at reducing the average number of car kilometres instead of car ownership - is also the one desired by experts, but in practice it is not expected that user behaviour will change drastically. For the other options the same rank order of scores as in the case of the expected options is found. Particularly the use of slow modes (walking and cycling) receives a much higher score. The use of clean and more economical engines is valued the same as in the expected case.

The option of changing spatial behaviour (mentioned by the respondents themselves) is given an average score of 8.64, which is striking when we see the low score of this option in the expected case.

Conclusions

The expected behavioural options are always valued lower than the desired ones. Hence it does not seem reasonable to expect that the behaviour of individuals will change as much as is deemed necessary by experts. It is only firmly expected that the use of cleaner and more economical engines will largely increase, but this is an option which influences the behaviour of users to the least extent. The other options - which have a larger impact on behaviour - are valued much lower.

The expected behavioural options are given moderate scores. This outcome confirms the scores in Section 3.9, indicating that individuals want to sacrifice very little in order to achieve a sustainable transport system. This holds for both reducing car use and increasing the use of slow and collective modes. It is clear that this inert behaviour may be a major obstacle for achieving a sustainable transport system.

The desired options are valued much higher. It would appear that a transport system is favoured in which the car is used much less, since reducing the number of car kilometres and car days is valued relatively highly. In Section 9.5 this will be investigated in more detail by discussing urban, regional and long-distance transport systems. On the other hand, the collective modes would be used much more than is expected.

9.2.2 Expected and desired policy measures

Expected policy measures

The results on policy measures are depicted in Figure 9.2. The only expected policy measure with an average score above 7 appears to be an increase in parking fees, which is already common policy in many Dutch cities; this policy is expected to continue. A second policy measure is an increase in fuel prices, which has also been introduced in the past decades. In the Netherlands a policy has also been announced for cutting down car parking facilities for firms as well as for city centres, an option which is given the next highest score.

Road pricing is also given a score above 6, so it may be concluded that the policy of the government is expected to be mainly oriented towards making car use less attractive. The existing policies will continue, but a system of road pricing might also be introduced.

Low valued, but less striking, is the reduction in car ownership. In the Netherlands, government policy aims at increasing the variable costs (i.e. car-use costs) and not the fixed costs (car ownership). No strong policy is expected in the field of offering better information about transport externalities. The road network will not become more diffuse. For urban transport, however, this option has been introduced on a small scale in several Dutch cities at the moment. It is apparently not expected that this trend will drastically change.

Desired policy measures

When we analyse the desired measures, the most striking finding is that the government should improve the alternatives for the private car, with a score of even above 8. Also the scores on the other desired measures are somewhat higher. Increasing parking fees and fuel prices gets about the same score. Road pricing now has a slightly higher score than reducing the number of parking places. Again the supply of adequate information is valued low; the same holds for reducing car ownership (which is just a little higher than the expected score).

Finally, making the road network more diffuse again scores the lowest, although the score is substantially higher than in the expected case. For the scores on the desired measures, the highest standard deviation is found for this measure; thus the scores given to this measure differ most.

Conclusions

When the expected policy measures are analysed, it appears that policies especially aimed at making car use unattractive are likely to be introduced, while investments in alternatives will lag behind. Especially policies which are already common (reducing parking places and increasing fuel prices), as well as introducing a road pricing scheme are the most likely policy measures to be introduced. It should be mentioned, however, that an increase in fuel prices is expected to have only a moderate impact on the use of the car and the use of collective modes.

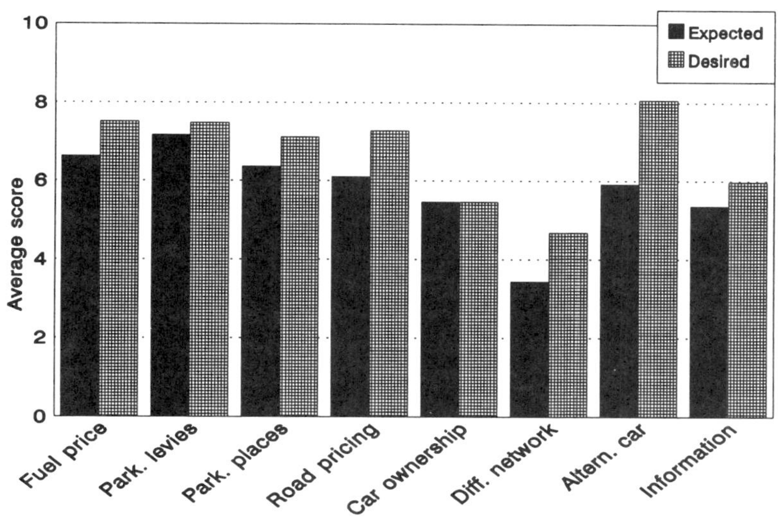

Figure 9.2. Expected and desired policy measures
Notes: 1) Scores 1-10; 1 is lowest, 10 is highest
2) n = 254-261

It is generally regarded as desirable, - next to making the car even less attractive - to invest much more in collective modes and new technologies. Apparently the priorities of the government are not expected to be high in this field, may be because of foreseen severe problems with government budget deficits.

9.3 EXPECTED AND DESIRED DEVELOPMENTS FROM A SPATIAL PERSPECTIVE

9.3.1 Expected and desired developments at the European spatial level and the resulting modal split

The expected developments at the European scale level

Transport and spatial organization of activities are two sides of the same coin. As argued in Part B, the spatial organization is of strategic importance for the need for transport and the future modal split. Therefore, expert views on the future modal split will in this subsection be linked to their views on the spatial organization.

At the European level the forum of experts expect that a moderate concentration of activities and population will occur in the core regions and the big cities (see Figure 9.3).

The spatial deconcentration trend may largely come to a standstill, or may shift to relatively new urban areas. On the other hand a large minority of the experts expect a development towards a moderate or entirely diffuse spatial organization.

The consequences for the modal split (between car and public transport) of these developments are that despite this moderate concentration the private car will gain a larger market share, while about 25 per cent think that the modal split will change in favour of collective modes and 20 per cent expect the modal split to remain about the same. Almost no one expects total mobility growth (or the growth of car mobility) to stop.

When a cross-table analysis is applied, it appears that respondents, who think that a more diffuse spatial development will occur, expect the use of the private car to grow faster than the use of collective modes, while a large minority expects only car mobility to grow. This may be explained by the fact that in a diffuse spatial organization many criss-cross transport links occur with a low density of demand, in which situation collective modes are not competitive.

When a mixed development is expected, the modal split will remain more or less equal. In contrast, when a concentrated organization is expected, the collective modes might increase or keep their market share, although a large majority thinks that even in this situation car use will grow faster than the use of collective modes. It is clear, however, that in a concentrated spatial organization, collective modes are expected to have more chances of gaining ground than in a diffuse spatial organization.

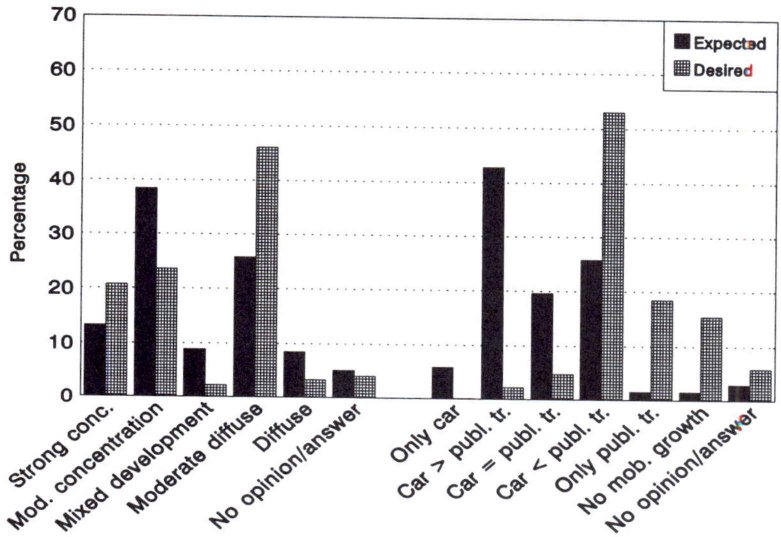

Figure 9.3. Expected and desired European spatial organization and the change resulting in the modal split

Note: $n = 271$

The desired developments at the European scale level

When the desired developments in Figure 9.3 are analysed, it appears that almost 50 per cent see a moderate diffuse organization as the ideal. This is quite surprising because this type of organization reduces the competiveness of collective modes, while the previous analysis indicated that these modes should be stimulated. Apparently the living conditions in a more diffuse organization are valued very highly, a goal which is thought to be more important than the consequences for the transport system and the wider environment. When the desired developments are compared with the expected ones, it appears that it is not desirable to have a further increase of the market share of the car or even to keep the modal split equal to the current situation. On the other hand, it is also expected by the experts that the use of the car will increase further. When a cross-table analysis is applied, it appears that this development in the modal split is supported by all the respondents whether they desire a diffuse or a concentrated spatial development.

Also interesting is the fact that about one third of the experts wants no mobility growth at all or only a growth in the use of collective modes. This group is clearly of the opinion that this is the most desirable way to achieve a sustainable transport system and put a low value possible advantages of more mobility.

Conclusions

It appears that at a European level a more concentrated spatial organization is expected, while a more diffuse spatial organization is desired. As may be anticipated, the expected modal split in a concentrated organization is more in favour of collective modes than in a more diffuse organization. In general, however, the collective modes are not expected to gain a large market share, not even with a moderate concentration of activities.

When we analyse the desired developments, a shift towards collective modes is desired, despite the associated less favourable spatial organization. So the barriers to sustainable transport systems are very high, even though this development may still be possible.

The expert views on the modal split are consistent with the conclusions on the issues discussed above: a system with a high share of the car in the modal split is expected, whereas a shift towards collective transport systems is desired.

9.3.2 Expected and desired developments at the urban level and the modal split

The expected developments at the urban scale level

At the urban level it is expected by 44 per cent of the respondents that there will be a (moderate) compact urban development, so that the policy to achieve more compact cities will largely succeed (see Figure 9.4). As a result the trend towards 'green suburbs' will largely stop. It is striking that only 14 per cent expect a more diffuse spatial organization, a share which is smaller than might be imagined, in view of the fact that the living conditions in diffuse cities tend to be better than in compact ones.

Next, it is interesting to assess how the expected spatial developments are related to the modal split. Most respondents expect the modal split to change in favour of the private car, while about one third thinks that the modal split may change in favour of public transport or that no increase in car use will occur.

When a cross-table analysis is carried out, it appears that only in a development towards a compact city a shift of the modal share may occur in favour of collective modes. So it could be concluded that only in a very compact city may the modal split change in favour of collective modes. However, the majority still expects the modal split to change in favour of the private car.

Desired developments at the urban scale

When the desired developments are analysed, a majority of the experts is in favour of a compact urban structure, while a large minority wishes a moderately compact organization. Only a small group thinks that a (moderate) diffuse spatial organization is to be desired.

When the resulting changes in the modal split are analysed, it appears that 94 per cent wish a shift towards public transport. Many respondents also favour a growth in car mobility, but a large group wishes no mobility growth or only public transport to grow. These findings are consistent with the findings presented before and with the desired spatial organization.

From a cross-table analysis it is possible to identify, in general, the same trend as was found for the desired developments and at the European scale level.

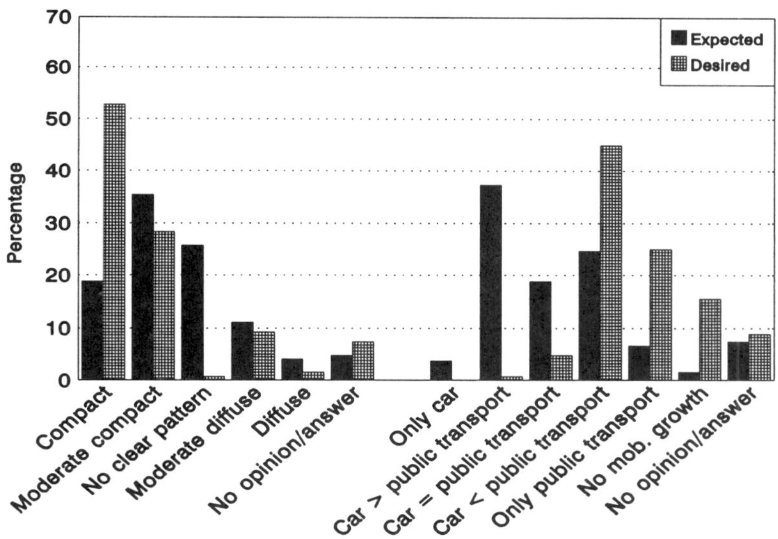

Figure 9.4. The expected and desired urban spatial organization and the resulting change in the modal split

Note: $n = 271$

Conclusion

It may be concluded that at the urban level a moderately compact development is expected, while a much stronger compact spatial organization is actually preferred. The expectations for the future modal split, however, differ much more. It is expected that the private car will gain a larger market share, despite the trend towards a moderate compact city. The desired developments, however, move in the direction of an increasing market share of collective modes. So this trend is by and large consistent with the previous findings.

9.4 EXPECTED AND DESIRED DEVELOPMENTS FROM INSTITUTIONAL, ECONOMIC AND SOCIO-PSYCHOLOGICAL PERSPECTIVES

Institutional factors

In the institutional field, experts were asked at which political level the most important decisions about transport are likely to be taken. The answers are presented in Figure 9.5.

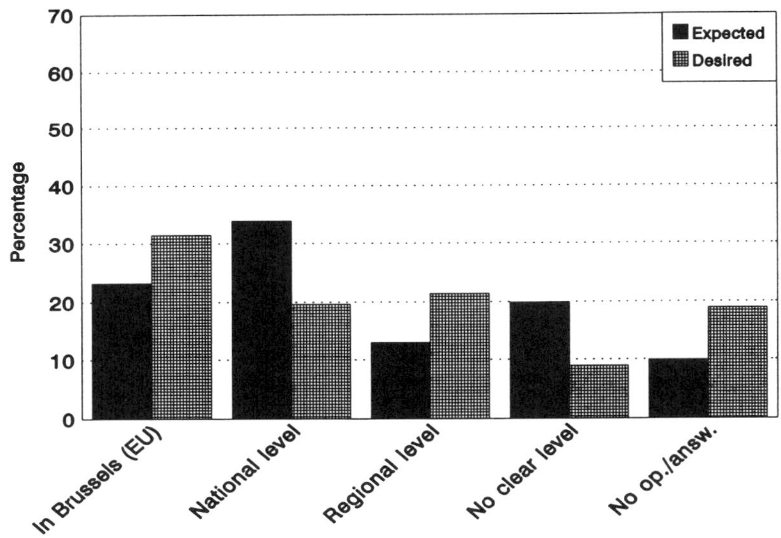

Figure 9.5. Level at which decisions are to be taken
Note: $n = 271$

It appears that the most important decisions are expected to be taken at a national level, while a large minority expects that the European Union (EU) will gain more influence. So it is expected that the current situation will also exist in the year 2030, but with more influence from the EU. The trend that the EU will gain much more influence is expected not to increase significantly, so that the

Member States will largely keep their national competence regarding transport policy and spatial policy.

When the desired developments are analysed, it appears that the influence on the national level should be much smaller than in the expected situation; the highest percentage is then found for the EU. Also the regional level gets a much higher percentage. A strong EU is apparently preferred, which may result in a more centralized transport policy, where standards and regulations may be set by the EU.

Economic factors

In the economic domain the respondents were *inter alia* asked to assign a 1 to 10 score on expected and desired possibilities for increasing the profitability of the railway companies (see Figure 9.6).

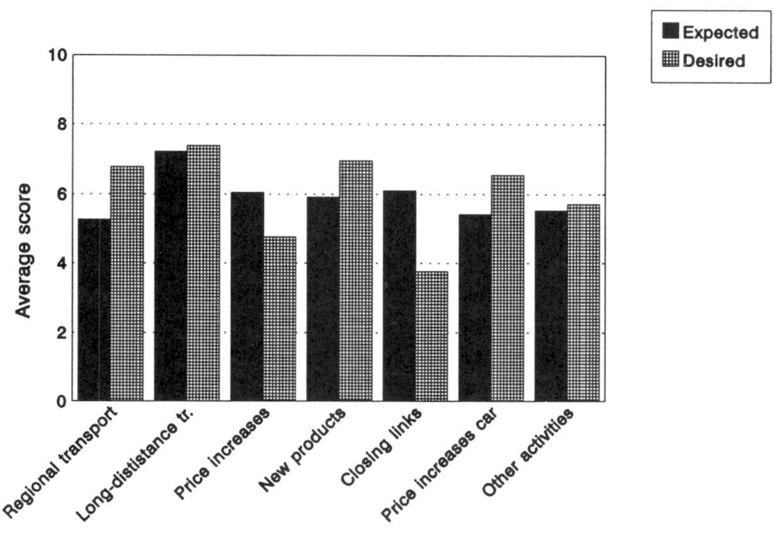

Figure 9.6. Expected and desired possibilities for increasing the profitability of the railways

Notes: 1) Score 1-10; 1 is lowest, 10 is highest
2) n = 228-248

When these scores are analysed, it appears that most profitability increases are to be expected in long-distance traffic, which was given the highest score. In most

European countries today plans are being launched to increase the profitability of long-distance traffic by making the operation of these links more independent of government influence. In general, this type of transport is already quite competitive with the private car; if long-distance traffic is more independent, profitability may increase even further. This may also happen by reducing the investments, for example, by making the network more diffuse. It will be clear that this may reduce the number of passenger kilometres which is expected to be reinforced by other measures.

For example, the profitability is expected to rise because of price increases and the closing of several unprofitable links. The railway companies in several countries (like the United Kingdom and the Netherlands) have already proposed to do this. Therefore, this may be the reason why these options are given quite high scores. The railway companies are not expected to be very innovative, since new products (e.g. HST) as well as other activities - the Dutch Railways are participating, for example, in a consortium for operating a second Dutch telephone network - are not expected to contribute to the profitability significantly. Finally, the lowest score is given to an increase in the profitability of regional transport, which runs at big losses now. It is not expected that this situation will improve very much in the future, a viewpoint which will only render support to the idea that the network coverage will have to be reduced.

In contrast to the expectations, when the scores for desired developments are analysed, entirely different scores are given. Long-distance traffic again has the highest score, which is not surprising.

Secondly, it is striking that the railways should develop new profitable products (like HST), which should contribute more to the profitability of the railways than is expected. It is interesting, however, that the expectations of experts on this option are fairly low.

Next, an increasing profitability of regional transport is expected. This can clearly only be achieved if regional transport is exploited in a more efficient way (maybe by introducing new products), which may reduce the need for subsidies. Also a higher score is given to making car use more expensive. A low score is once more given to complementary commercial activities, so that according to the experts the railways should not invest very much in non-transport-related activities.

It is noteworthy - especially when we see the expected developments - that very low scores are given to price increases of rail transport and reductions of the network. Apparently, the use of railways should be stimulated by avoiding charging too high fares and keeping unprofitable links open.

In conclusion, the experts pessimistically expect in general that the railways will increase their fares and reduce their network without offering innovative new products, while quite the opposite policy is regarded as desirable by these experts.

Socio-psychological factors
The majority of respondents expect the individualization trend to continue at about the same pace as at present (see Figure 9.7). In Section 3.9 it was analysed that this might have quite negative impacts on the efforts to achieve a sustainable transport system. It is not surprising, therefore, that according to the majority of respondents this trend should be stopped or reversed. It should be mentioned, however, that a large group had no opinion about the desired development or did not give an answer, since the individualization trend has many - as yet often unknown - consequences in other fields outside transport.

It appears to be desirable to reduce social and recreational traffic. It appears that about 50 per cent of the respondents would like a large or very large reduction in social and recreational traffic, which is a surprisingly high share, especially because it is very hard to stop this growth. It may, therefore, be concluded with some caution, that growth should be restricted to 'economically meaningful' traffic, because apparently transport causes so many externalities, that a clear economic profit for society at large should be generated in order to justify this. In the Netherlands, for example, it is being discussed whether to give business (and goods) transport priority on the road network above other kinds of transport.

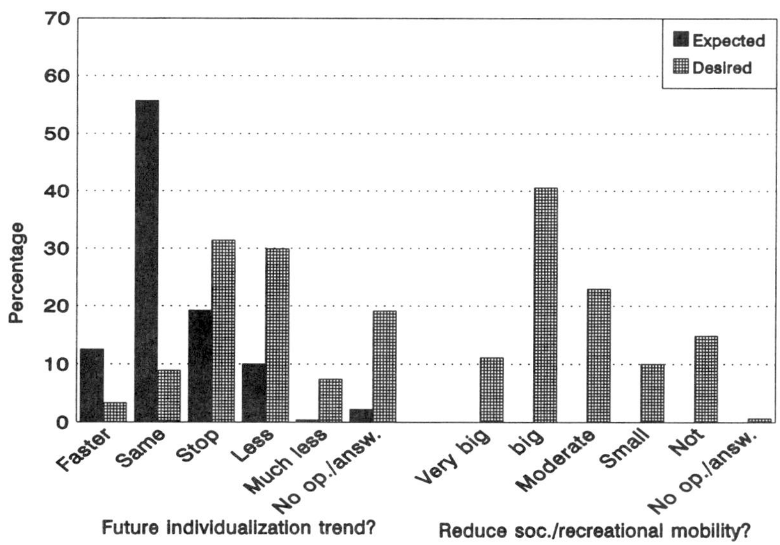

Figure 9.7. The expected and desired trends in the socio-psychological field

Note: $n = 271$

Conclusions

From the answers given on the questions in these distinct fields, it is evident that the desired scores differ largely from the expected ones in almost all cases. Most trends will continue or remain about the same, while major changes are desired by experts. In general, it may be concluded that the expected scenarios will be quite different from the desired scenarios. From the scores given to possible increases in the profitability of the railway companies, it appears that the railways are expected to introduce price increases and to close down parts of their network. The desired situation is quite the opposite, however. From this observation we arrive at the paradoxical conclusion - common in many countries - that the railways may be forced to reduce their activities, even though people really want to stimulate the use of these services.

9.5 THE TRANSPORT SYSTEM IN 2030

After the general analysis of expected trends and desirable developments, we will now focus on a more detailed analysis of the composition of the transport system in 2030. In Sections 9.2-9.4 some options and measures have been discussed in general, in the present section we will focus on subsystems and the expected and desired introduction of new transport modes. As mentioned above, the expert views on the transport system in the year 2030 have been subdivided into three submarkets: urban (less than 20 km), regional (20 to 50 km) and long-distance transport (more than 50 km). These have in turn been considered in relation to three categories of impacts:

- the importance of several policy measures;
- the importance of conventional transport options;
- the introduction of new transport modes and technologies.

The respondents were asked to value all options and modes by means of scores on a 1 to 10 scale (which is the usual method of valuing items in the Netherlands) for the expected as well as the desired situation. The measures discussed here were already dealt with in general terms in Section 9.2.2; now some of the measures will be analysed per distinct submarket. In this way the transport system in the scenarios can be identified and assessed in more concrete terms here than was possible on the basis of the more global analysis in previous sections. Also the characteristics of the three submarkets will be discussed in greater detail.

9.5.1 The urban transport system in 2030

Expected and desired policy measures

The results for the expert views on future urban transport are given in Figure 9.8.

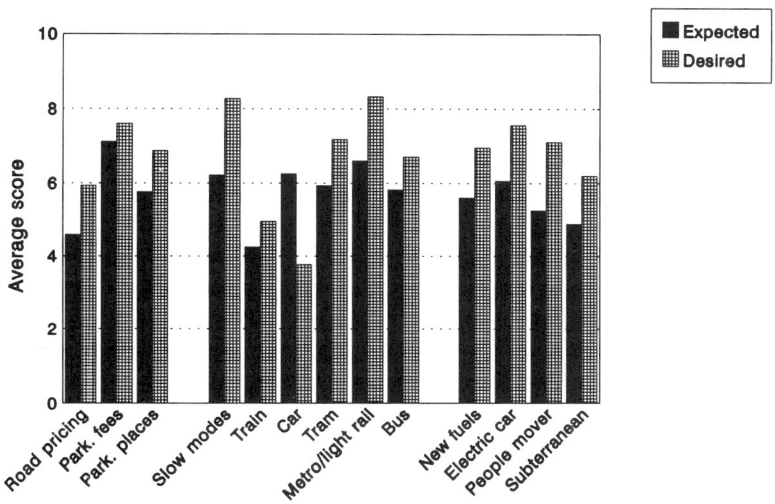

Figure 9.8. The expected scores on measures and modes in the urban transport system

Note: 1) Score 1-10; 1 is lowest, 10 is highest
2) n = 254-266

The highest score for expected policy measures is found for an increase in parking fees which may make the use of private cars less attractive. The second highest score is given to reducing the number of parking places, which is expected to be introduced to a somewhat lesser extent than an increase in parking fees. The same rank order was already found in Section 9.2.2. Road pricing (or in cities maybe cordon pricing) is not expected to be introduced on a large scale, given that the score is far below 5.

The scores for the desired measures are all higher than for the expected ones, although for the parking fees this score is only just a little higher. It is still the highest score, however, with a reduction of the number of parking places in second place.

The new technology of road pricing also gets a much higher score, viz. above

6, so it may be concluded that this option is also desired to be introduced on a reasonable scale. As might be expected, the standard deviation of this measure is the highest, so that the experts disagree most on this measure.

Other policy measures which are mentioned by the set of respondents are: changing the spatial organization (8 times, scores 4.9 (expected) and 7 (desired; see also Section 9.3), park and ride systems (once) and car sharing ('call a car', 3 times).

Expected and desired use of conventional modes

The scores by the experts on the expected use of conventional modes all fluctuate around 6 (except for the train). The highest score is assigned to metro and light rail systems, which are expected to be introduced on a larger scale than is the case now. This is a quite likely situation, since new links are under construction in several Dutch cities (Amsterdam, Rotterdam). For the remainder, the car and slow modes (cycling, walking) are given the next highest scores, while the bus and tram get somewhat lower scores. It seems plausible that the score for the train is somewhat lower, since this mode is not primarily meant for urban transport, although the train is sometimes also used as a mass transit system.

When the desired scores are analysed, it appears that all modes gain higher scores, except the private car, which is plausible because of the high external effects of this mode. The slow modes and the metro/light rail, in particular, now get very high scores of above 8, implying that the modal split in urban traffic should change in favour of these modes. The tram and bus also get relatively high scores, while the score of the private car is even below 4, so that a much lower modal share is regarded as desirable for the conventional car.

Expected and desired use of new modes and technologies

Respondents were also asked to assign scores for several new technologies and modes which might be introduced up to the year 2030 (see also Chapters 5 and 6). The highest score is found for electric cars, although this score is only about 6, so that only a limited introduction is expected. The second highest score is found for new fuels, which means again an improvement of the position of the private car. The people mover is given a rather low score, while subterranean transport is expected to be introduced only on a small scale. So especially improvements of the private car are likely to be introduced, while the other developments have far less chances in the opinion of experts.

The desired scores are all higher than the expected scores. Thus, it is desired that more of the new modes and technologies should be introduced than is at present expected. The electric car again has the highest score, followed by the people mover. Also somewhat higher - but still low - scores are given to the introduction of new fuels and subterranean transport by car and train. Other modes mentioned by the experts are passenger transport by water, improved human-powered vehicles and the electric bicycle.

Conclusions

From the scores given for the urban transport system it may be concluded that the private car and its improved versions are expected to dominate the urban transport system in future, although various policy measures will be introduced to make the car less attractive. These results are consistent with findings for other issues on the future development of the modal split. In Section 9.3.1, it appeared that these developments are expected to occur in combination with a moderate compact spatial organization of our society.

In the desired situation, however, there is more emphasis on collective modes, while also more and severe measures should be introduced to make the car less attractive as the dominant mode. These developments should be favoured in a situation where the urban spatial organization becomes very compact. Again these results are consistent with expert views on developments in these fields.

9.5.2 The regional transport system in 2030

Introduction

The regional submarket is only given modest attention in this study, because the focus is on urban and long-distance transport. This choice is made mainly because in this submarket the least number of new developments are to be expected. Nevertheless, it is still an important one, with its own specific problems. The average scores given to the various options are presented in Figure 9.9.

Desired and expected measures

The highest score - but still a relatively low score (below 6) - in the regional submarket is given to road pricing, while car pooling gets an even lower score. It may be concluded therefore, that it is not very likely that either of these measures will be introduced. The corresponding scores in the desired case are much higher; both options should be introduced on a rather large scale. Other options mentioned are a change in the spatial organization and improved park-and-ride systems (people drive to the borders of urban areas by car and change there on to high-quality urban transport).

Expected and desired use of conventional modes

It is not surprising that the highest value is given to the private car, despite the high external costs of this transport mode. Other rather high scores are assigned to train and metro/light rail systems both with scores above 6. Thus these modes may also be important in the regional transport system.

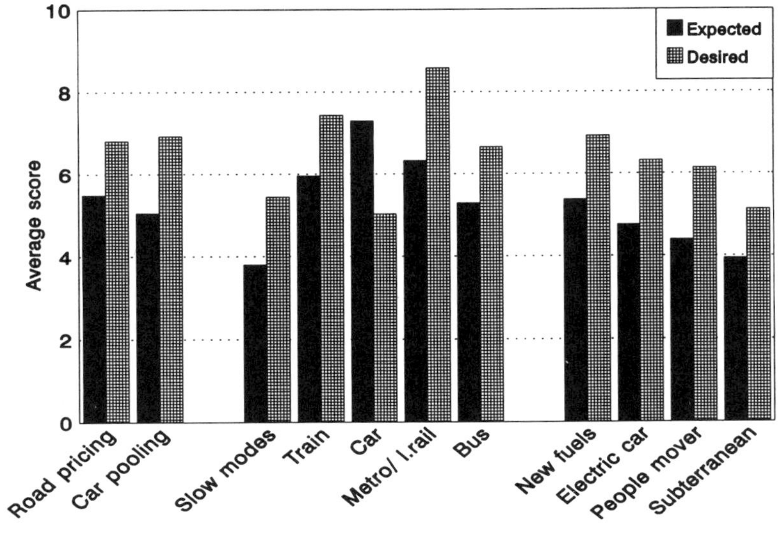

Figure 9.9. The expected scores of measures and modes in the regional transport system

Notes: 1) Score 1-10; 1 is lowest, 10 is highest
2) n = 252-261; for metro/light rail n = 23-25

The bus is expected to be less important in the modal split. Hence, regional bus transport is not expected to be of major importance, although this system is quite important nowadays in a country like the Netherlands. The average distances are apparently too long to be covered by walking or cycling, so a very low score is given to the slow modes.

When the desired scores are analysed, an entirely different picture is found, especially because the private car gets the lowest score. This is not surprising because of the high externalities of this kind of transport. The metro/light rail systems are getting by far the highest score now. Also high scores are given to the train and to a lesser extent the bus. Slow modes get relatively higher scores, too, although the scores for these modes still remain relatively low.

Expected and desired new technologies and modes

All expected scores for the new modes and technologies are quite low, so that large-scale introduction is not very likely. New fuels receive the highest score, which is a higher score than that for electric cars. This is striking since electric cars may also drive longer distances in the future, but it is apparently expected that conventional cars used for regional traffic will also be used for long-distance traffic (which is not possible at present with electric cars because of their low action range). In the urban transport system, the electric car gets a higher score than new fuels, however. Low scores are given to the people mover and subterranean transport, so that a large-scale introduction of these modes is not expected.

The desired values appear to have the same order; only the scores are much higher. New fuels, electric cars and the people mover all have scores higher than 6, so these technologies and modes should be introduced on a rather large scale. Subterranean transport, however, is not immediately seen as a desired option.

Conclusions

It may be concluded that the car and its improved versions are expected to dominate the regional transport system in the future. There will apparently be no good alternatives. Moreover, the expected policy measures to reduce car use will not be introduced to a great extent.

In the desired scenarios, the metro and light rail systems especially should be introduced on a large scale. The same holds to a lesser extent for new fuels, the electric car and the train. The conventional car should largely lose its modal share.

In general, it appears that it is even more difficult to change the regional transport system than the urban system, since the scores given for the alternatives for the car - besides those for metro and light rail - are relatively low.

9.5.3 The long-distance transport system in 2030

Expected and desired measures

For the long-distance transport system road pricing gets a score which is lower than 6, so that introduction is not very likely (see Figure 9.10).

So in all distance categories road pricing is not expected to become very important. Car pooling, however, gets a score even below 5, and it is thus not expected that car users are prepared to give up their individual transport behaviour. The score is even lower than that for the regional transport system.

The desired measures are all given higher scores, similar to the case with the other submarkets. The scores are now higher than 6, so that it may be concluded that these options should ideally be introduced on quite a large scale. Other options mentioned are again a change in the spatial organization and new automated systems of car sharing.

Expected and desired use of conventional modes

As in the other submarkets, the private car still gets the highest score. Therefore, it is not expected that the other modes will capture a high market share from the private car. It is striking, however, that the train also has a very high score for this distance category, so this mode is expected to be an important one in the future.

The plane gets a rather low score, despite its high growth rates in the past decades. Finally, the bus gets the lowest score, although this mode is an important one for long-distance trips in several countries. This low score may be influenced by the fact that the bus is not a very important mode in the Netherlands for long-distance transport. Anyway, it is expected that the bus will lose a significant part of its market share.

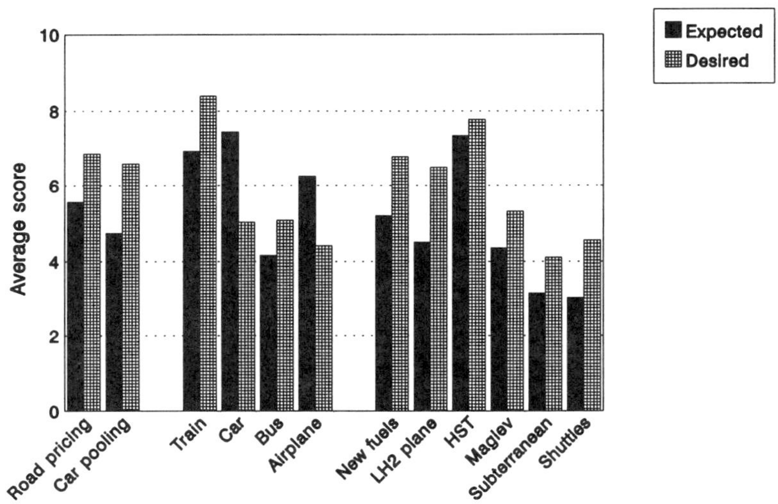

Figure 9.10. The expected scores of measures and modes in the long-distance transport system

Notes: 1) Score 1-10; 1 is lowest, 10 is highest
2) n = 214-262

Clearly, the desired scores differ considerably from the expected ones, since the modes with high externalities (car and plane) get a much lower score. These modes should become less important in the view of experts, but the expectations that this will happen appear to be low. The train gets a very high score (above 8) now, so that this mode should ideally become more dominant in the future. Striking again is the low score for the relatively clean and cheap bus.

Expected and desired new technologies and modes
As discussed in Part B, there are many new technologies developed for long-distance transport. The HST is the mode which is planned to be introduced on a large scale, so that it is not surprising that this mode gets the highest score of all new modes and options.

The other modes and options get a much lower score. New fuels for the private car are given the highest score in this category, which may be due to the same reasons discussed above for the other submarkets. The average score is relatively low.

Next, the Maglev high-speed train is identified as a new option; the score is far below 5, however. The same holds for the LH_2 plane, subterranean transport for conventional transport and shuttles through vacuum tunnels. It should be mentioned that many respondents did not give an answer for the LH_2 plane, because they did not feel able to assign a score for this option. None of these options is expected to be introduced on a large scale.

The desired scores are higher for all new modes and options. The highest score is again given to the HST, although this score is not much higher than the expected score. Next, new fuels stand out with a much higher score; the same holds for the LH_2 plane. Subterranean shuttles now get a slightly higher score than the Maglev, but both scores are still low. The same holds for subterranean transport. Thus, none of these modes are seen as desired options to be introduced on a large scale.

Conclusions
It may be concluded that the private car is expected to stay an important mode in the future, but that the conventional train and the HST may also become important modes in the long-distance transport system of the year 2030. The modal split may therefore be more in favour of the collective modes. In the desired system the conventional car and plane have a much smaller market share, while new fuels, the train and the HST, especially, have a somewhat larger share. Again the findings are that the modal split should be more in favour of the collective modes and new technologies than in the expected situation.

9.5.4 Conclusions about the transport system in 2030

When the overall scores are looked at, it appears that in general a transport system for the year 2030 is expected in which the private car, and to a lesser extent, its improved versions, will remain the dominant modes. This holds true particularly for the urban and regional transport system, while in the long-distance transport system the train and HST may have a rather high modal share. Measures to reduce the use of cars are not expected to be introduced to a great extent.

In contrast, in the desired system the use of the conventional car is much smaller; when a car is used it should be run with clean fuels or electricity. It is considered that collective modes (in particular metro/light rail and the train) deserve a much larger market share.

In the next chapter the results presented here will be used to construct expert-based scenarios by applying the Spider model discussed in Chapter 8.

10 NEW TRANSPORT FUTURES DESIGNED BY EXPERT OPINION

10.1 INTRODUCTION

Scenario design presupposes two conditions: a frame of reference and proper information in order to evaluate alternative futures. In Chapters 7 and 8 an introduction was given to the construction of scenarios in general and the methodology used in our study, based on the Spider model. In the present chapter we will construct expert scenarios based on the Spider approach. The scores for the attributes (background factors) of the various scenarios - in the form of rank orders on the eight axes of the Spider's web - will be assigned and derived from expert opinions obtained by means of the survey questionnaire among Dutch transport experts (already detailed in Chapter 9). Clearly, the underlying assumption is that these experts have more insight into the feasibility and desirability of new transport systems than anyone else. As a further development of the methodology, these views and the resulting scenarios have been retrospectively re-evaluated by these experts on their plausibility, consistency with policy objectives and sustainability. To complete the research project, the scenario outcomes have also been re-assessed at three international workshops by European transport experts, with the help of a (brief) questionnaire.

First, we will describe how the expert-based scenarios were constructed with the use of the questionnaire. Then the design of both the expected and the desired scenarios will be discussed. A brief general introduction to the successive scenarios will be given, then the future pictures emerging from the scores on the axes are analysed, and finally the resulting transport system in 2030 will be presented and discussed in greater detail. To conclude, a judgement (or validation) of the scenarios will be offered, making use of the above-mentioned self-assessment procedure of Dutch experts and the re-evaluation round among European transport researchers.

10.2 THE CONSTRUCTION OF EXPERT-BASED SCENARIOS

The Spider approach is used as the basis for acquiring and expressing views by a representative sample of researchers. The transport experts were asked to give their views - by means of scores or rank orders - on expected and desired trends (see Chapter 9). Their aggregate views - based on the maximum frequency of any one item (mostly on a qualitative rank order scale ranging from 1 to 5) - were then depicted in the Spider model. By using these opinions, it is then possible to construct an 'expected' and a 'desired' scenario. Clearly, not all aspects of future developments could be investigated in full depth, in order to keep the questionnaire within reasonable limits. This means that in designing the various scenarios some consistent interpretative reasoning had to be followed, among others with the help of participants of various international workshops devoted to the topic of this study. In other words, the scenarios are elaborated opinions and interpreted views of the respondents, supplemented with opinions of participants in the international workshops. As mentioned above, a second (short) survey has been used as a check for the assessment and validation of the scenarios.

For most items of the scenarios the majority answers - or the average scores - were taken as the starting-point for the analysis, while in relevant cases the answers of significant minorities were taken into account, too. The items used for the scores on the axes are presented in Figure 10.1.

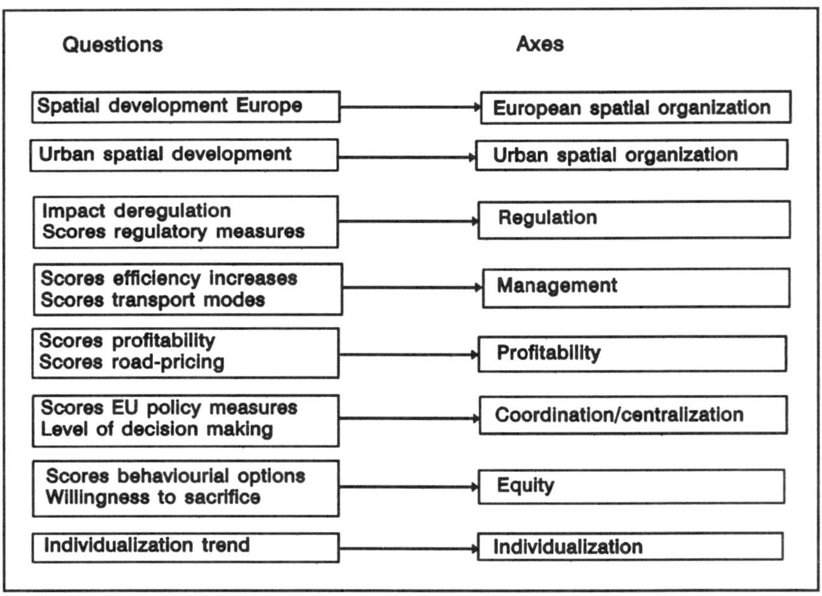

Figure 10.1. The translation of survey questions into scenarios

It may be clear that the distinct axes are not entirely independent from one another, a situation which is impossible to achieve. The definition of the items on the axes has been made as independent as possible, however, and we will concisely recapitulate them here.

Axis 1 of our Spider's web is based on the expert opinions on the *European spatial organization*; for Axis 2 the same has been done for the *urban spatial organization*.

In the institutional field the opinions of experts on the impact of the current *deregulation* trend are used, together with the scores on regulatory measures (*vis-á-vis* market-based measures). For example, when most market measures (e.g. fuel price increase, road pricing) get higher scores than regulatory measures (e.g. restricted car ownership, reduction of parking places), it may be concluded for the purpose Axis 3 (regulation) that overall the intervention of the government is more market-oriented.

On the *management* Axis 4 the emphasis is on efficiency increases (within the context of profitability increases and policy measures). When these increases largely depend on government policies, it is assumed that the management of transport modes and transport infrastructure will largely be the responsibility of the public sector. Also the scores on the management of several transport modes have been used to calculate the scores on Axis 4, since we have seen in Chapters 5 and 6 that some modes are more dependent on government influence than other modes.

In the economic field the scores for the profitability of the railways and the scores for road pricing (as a direct price for the use of private cars) have been used as indicators for the expected and desired score on the *profitability* Axis 5. On Axis 6 the indicators for the level of *coordination/centralization* are found by using scores on policy measures and the policy level at which they should preferably be introduced. For example, a large-scale fuel price increase needs coordination, because otherwise evasion behaviour will occur by buying fuel across the national border and illegal imports. Also the expected and desired level at which the most important decisions are taken have been used for these respective scores on Axis 6.

In the socio-psychological field the scores for the behavioural options have been analysed from the viewpoint of *equity* - Axis 7 - some options may be expected to lead to more equity than others. Also the willingness to make sacrifices to achieve more sustainability is used for determining the scores on this axis. The question on how the *individualization* trend is expected and desired to continue has been used for Axis 8.

The *final picture* of the resulting transport system has been determined by using the expected and desired impact of the spatial organization and the scores on the distinct transport modes in the various submarkets of transport.

Although the different axes in the Spider model may certainly impact on each other, this way of designing scenarios ensures that as far as possible the scores on the axes can be identified independently from one another. In the next sections we will describe the design of the resulting 'expected' and 'desired' expert-based scenarios in greater detail by using the results of the survey in terms of the eight axes discussed above.

10.3 THE EXPECTED SCENARIO

In presenting the scenario results, we will first describe the *expected* scenario. When the answers given by experts are analysed, and if we look at the highest-scoring items, it may be concluded that many current trends will to a large extent continue, while the governmental policies and measures are less influential than was intended: in sum the expected developments are less extreme than in the reference scenarios. In this respect, it is interesting to note that more than 80 per cent of the respondents answer 'may be' or 'no' to the question about whether the transport system will be sustainable in the year 2030. As a result the private car will largely keep its modal share, while collective modes and new technologies are introduced only on a rather small scale and in an incremental way. A technology push in the transport sector, characterized by radical innovations, is not regarded as plausible. If this scenario is to result in a sustainable transport system, a drastic improvement of the private car is necessary, otherwise a sustainable transport system (e.g. a large-scale reduction of CO_2 emissions) cannot be achieved (see also Chapter 12). From the answers, presented in Chapter 9, the scores on the various Spider axes may be drawn, as indicated in Figure 10.2.

At first sight the expected scenario is largely an extrapolation of current trends, although several behaviourial changes and policy measures may occur. These are not as big and severe as in the desired situation, however. Therefore, the resulting transport system also does not differ significantly from what is present.

In the next sections the expected scenario will be constructed by describing the distinct scientific fields and axes, in the same way as the reference scenarios have been discussed. As an end-stage the transport system which emerges from the answers and scores given will be presented.

10.3.1 Spatial factors

The European spatial organization

The spatial policy of governments is expected to be focused on concentrating activities and population in the large urban areas. In this way the externalities caused by transport are expected to become lower, because of the improved competitiveness of collective modes and the smaller travel distances.

Most probably, the suburbanization trend will largely be reversed as a result of this policy. Some -especially wealthier - people, however, will prefer not to return to - and to live in - these cities. As a result, they will prefer a residence outside these areas. But most people will adhere to the policy of the government;

therefore, a moderate concentration of activities and population will occur at the European spatial level. This will be the case in the core regions particularly, because already most people live there, while economic growth will be the highest there.

Given these expectations, several metropolises will appear to be most attractive for business firms and people, because of the spatial structure and agglomeration advantages. In these metropolises many head offices of international companies, financial markets, large industries, junctions of HST connections, an important mainport, etc. will be found.

These metropolises will be less dominant than is assumed in the specialization and concentration model, however, mainly for two reasons.

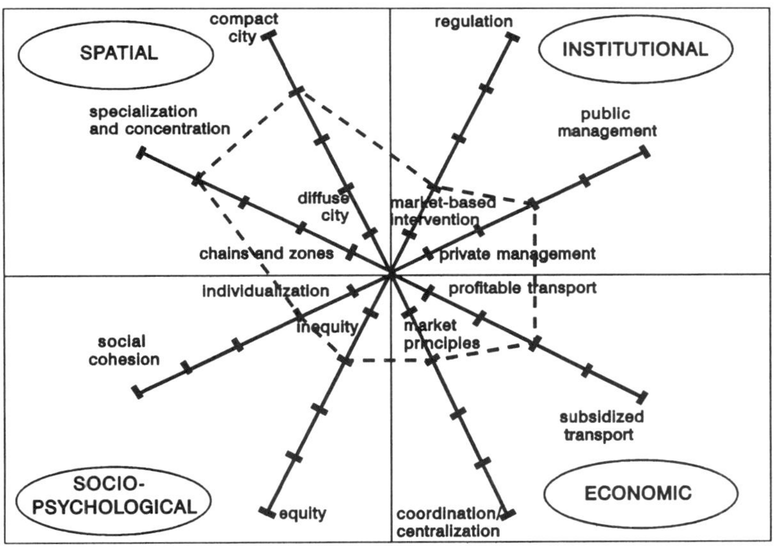

Figure 10.2. The Spider model: the expected scenario
Note: $n = 271$

In the first place, the concentration policy will not be entirely successful, as mentioned above. Therefore, several activities and part of the population will stay in the rural green regions. The second reason is that the European integration process will slow down or may even largely stop, so that the national states are expected to remain the most important political competence level. As a result, much governmental support will be given to the development of Europoles, especially in those countries in which no metropolis is found. In almost every country at least one Europole will be developed (in most cases the capital city). These are major centres of activities and of regional and national services, for example, departments of governments, an HST-terminal, a harbour and/or an airport.

At a lower level in the spatial organization, smaller cities will be found in which mainly regional activities are concentrated, but they will still be rather important for the economy. This spatial development is expected to have negative consequences for the rural and peripheral regions and economies. In general, only the highest income group will be able to live outside the cities.

In conclusion, the spatial organization will in general develop towards the specialization and concentration model, but not entirely. The metropolises will be less dominant than in this latter model, while the Europoles - and to a lesser extent other cities and rural regions - will become relatively more important.

The urban spatial organization

The cities in the expected scenario are expected to evolve into a moderate compact spatial structure, which will be a result of the official governmental policy. This will be imposed in a strict way in order to reduce externalities. In this urban spatial organization, most activities and the population will concentrate in the city centre and several subcentres. In these centres main nodes of public transport (e.g. HST-stations), offices, shopping malls, etc. will be found. Some of the population and certain companies, however, will locate in other sites, mainly because housing and space will become rather expensive in these centres and because of the better living conditions outside these compact cities. There will also be many housing problems, as a result of the population density, the larger number of households and the continuing individualization. Therefore, the urban centres will be too small to house all people and activities.

The lowest income groups will mainly reside in the old quarters of the cities, because they cannot afford to live in the expensive city centres. In these districts housing will be relatively cheap, but also of low quality. These quarters will not be very accessible either, because only low-quality public transport will be available.

10.3.2 Institutional factors

Intervention by the government

Generally speaking, in the expected scenarios devolution trends will continue, although governmental influence will still be rather high in several fields. Most emphasis will be put on measures to keep the national economy competitive with other European countries and other trade blocks as well as with ‘low wage countries’. For example, social security systems may largely be abolished.

Average tax rates will be much lower than nowadays, because this is assumed to be the best way to make the national economy more competitive.

The development of the European Union as a more influential authority will largely stop, since the Member States and the population want the national states to be the main political level. But in some fields, where cooperation is profitable for all countries (e.g. trade policy, several environmental measures), the EU will gain more influence.

In most countries a strict spatial concentration policy will be introduced, to reduce externalities caused by a growing population (for example, to save natural areas), transport, etc. For the remainder many policies, like regional development and agricultural policy, will differ for each European country. In general, however, these policies will largely be abandoned or at least significantly reduced.

Organization of the transport market

Transport is expected to remain a key responsibility of the national governments, and hence many of today's policy aims and trends will continue; current measures (e.g. increasing fuel prices and parking fees, autonomy of the railway companies) will still be the most important ones. In the EU, strict standards about emission levels of the private car will be imposed, which will to a great extent reduce the emission levels of the private car. Standards for telematics and physical route guidance will be set by the EU, too.

The railway companies will be made more independent of the governments, while the governments will mainly become responsible for the construction of infrastructure. The governments will not invest heavily in the collective modes, however, because of lack of funds. This is one of the reasons why these will not succeed in winning a higher market share on many links.

In most countries the railway companies will be divided into sectors per submarket. Long-distance transport will in most countries be profitable since in this submarket the train and its improved versions will be competitive with the car. This profitability will be achieved to some extent by increasing the prices and closing unprofitable links, while efficiency gains (e.g. by automation) will also make a major contribution to the improved profitability. The rise in fares of collective modes will be higher than that of the rise in car costs. Moreover, because of the favourable spatial organization, cars are still expected to stay attractive for many trips, they will, therefore, keep their modal share on many links, but in total their market share is not expected to become much higher.

Collective regional transport will become more unprofitable and remain subsidized by the government to a somewhat lesser extent than at present because of the closing of unprofitable links and price increases.

An attempt will be made to make the use of the private car less attractive by increasing the fuel prices and by introducing road pricing on the most congested links. In cities parking fees will increase and the number of parking places will be reduced. These policies will not become very popular, however, because people simply do not want to give up their mobility and individual travel behaviour, and collective modes will often appear not to be an attractive alternative. As a result, these policies will not be introduced to any great extent. Therefore, the government will stimulate research into making the car more economical and may stimulate the introduction of electric cars in urban areas, where their low action radius is not a big problem.

10.3.3 Economic factors

Development of the European economy

The European countries are expected to largely integrate their markets, but further cooperation will not be achieved. The national states will compete with each other and other parts of the world, a situation which will have negative impacts on the coordination of, for example, environmental policy. Governments will largely withdraw from many sectors and fields in society, in order to reduce taxes and to achieve competitive advantages. Thus, more emphasis will be put on introducing market principles in the economy and reducing regulations. This will also be necessary because of the continuing individualization in society. As a result, the European economy will develop positively, and will keep its relative economic position in the world *vis-à-vis* the United States and Japan, although the European Union will lose its position relative to several other countries, especially in the Pacific region.

This, together with the policy focused on spatial concentration, will have rather negative economic impacts on rural and peripheral regions. The European core regions will develop relatively favourably. These regions may appear to have the most competitive economy, and this trend will not be curbed by, for example, regional support from the EU, since this policy will be largely abandoned or at least reduced.

Profitability of the transport modes

The government is expected to partly withdraw from the transport market. Long-distance transport by railways will mainly be left to the market, and will become profitable on most links. Many efforts will be made to increase the profitability of regional transport. It would appear, however, that this will only be possible by closing many links. This will cause so much resistance in society that this policy will have to be ditched.

For urban public transport, ticket prices will be raised to some extent. The moderate compact structure of the cities will stimulate the profitability of urban transport, but subsidies will remain necessary to operate these modes.

Car transport, on the other hand, will be left relatively free of most charges, because such a measure would not be well accepted in society. The only measure imposed on private cars would be the introduction of strict standards on emission levels, etc., which will largely reduce the externalities.

Privatization of infrastructure will not be possible, and therefore infrastructure provision and financing will remain the responsibility of the national governments. Road pricing, for example, is hardly likely to be introduced, while the rail infrastructure also will not be profitable in most cases.

10.3.4 Socio-psychological factors

Individualization

The individualization trend in society will continue at the current pace. As a result, a society with more and smaller households will develop, with an increased emphasis on individual modes, while social cohesion in society will largely disintegrate.

This breakdown of society is expected to partly obstruct the spatial policy of the government, because more individual dwellings will be needed, while the planned change towards collective modes will not occur. Although it will be widely accepted in society that environmental measures are necessary, it appears that individuals will often not be prepared to make financial sacrifices in order to achieve a more sustainable society. They will not want to give up their individual travel behaviour (by car), for example. Notwithstanding this, people will nevertheless also be prepared to buy more expensive and cleaner cars, which in any case will be compulsory because of standards set by the EU in order to reduce the externalities caused by transport.

Equity in society

The government will significantly reduce its social security system; the same will hold for many subsidies, such as some of those for public transport. As a consequence, inequities - measured, for example, by income distribution - will increase. It will be accepted as a necessary policy to lower taxes in order to improve the competitiveness of the national economies.

In general, three economic groups may be distinguished. The highest income earners will account for about 10 to 20 per cent of the population. These people can afford to live outside the cities or in green suburbs and will own a private car, which is used for most of their transport needs.

About 70 to 80 per cent of the population will belong to the middle-income group of earners. This class will mainly live in the centres and subcentres of the compact cities. They will own a private car, which is run on conventional fuels or on electricity. The latter type of car will be cheaper than it is today, because this technology will be mass-produced, and there will also be some government support for this mode. It will still have a low driving radius, however, and it will therefore be used by people who almost never travel longer distances. Because of the moderate compact structure of the cities, the urban population will mostly live near nodes of collective modes, which will tend to make these more attractive in the congested urban areas and in other areas with high density transport links. Therefore, these modes will be popular for travelling between the compact centres of the larger urban areas, especially for the people who only have an electric car.

Finally, the lowest income group will account for 5 to 15 per cent of the population. These people will mainly live in the old living quarters of the large cities, where housing is cheap but of low quality. Such people will not be able to afford a private car and therefore will be dependent on the services of collective modes. High-quality nodes (HST and intercity stations) will be a bit remote, and will have to be reached by low-quality modes like bus or tram. These will be rather expensive for this group to use, and so their mobility alternatives will largely be reduced, which again will increase inequity in society.

10.3.5 The transport system in the expected scenario

Long-distance transport

In general, a radial transport system is expected to emerge with rather high-density links between the metropolises and between the metropolises and Europoles. On these links HST connections will offer a high-quality service. These services will mainly be used by people who travel to and from the urban (sub)centres. Especially in the core regions, these will have a somewhat high modal share, because here the transport links will have a high density of users, which makes them profitable.

Most people, however, will still use the private car for long-distance transport. These cars will mainly use fossil fuels, but much more economically. These reductions will also be achieved by intensive electronic guidance of vehicles. New fuels will not be introduced to a great extent, because they appear to have too many negative consequences for the travel behaviour of individuals, while conventional engines are much cheaper and can be technically improved.

Especially over very long distances planes will offer a good service, but this will be made more expensive by imposing several new taxes. These will be introduced because of the environmental resistance in society towards this mode, which results from the many negative externalities of aviation.

Subterranean transport is expected to be too expensive and people may be afraid of using the long tunnels, so subterranean infrastructure will only be constructed where absolutely necessary. Also the Maglev trains and subterranean shuttles will not be introduced on a large scale, because of their high costs and the fact that they are not compatible with conventional modes.

To connect the smaller cities, slower trains will be operated; the same holds for the main links in regional transport. Again, however, most regional transport will be covered by private cars, since the demand on these railway links is too low.

Mobility growth will largely continue, but its growth rate is expected to be lower than in the 1990s, since spatial concentration will reduce travel distances. There will not be much construction of new infrastructure, however, because the existing road infrastructure will be used much more efficiently by guiding cars, and by telematics or physical guidance especially on the main links. Another reason for this is that government budgets will not be sufficient for large-scale investments, and in particular not on the infrastructure for collective modes.

The modal split for long distances will remain in favour of the private car, which will account for about 70 per cent of the transport demand. Collective modes will account for some 30 per cent of the modal split; in particular their modal share in transport between the metropolises and Europoles will be relatively high being about 50 per cent.

Urban transport

In the cities a radial transport system will be established with main links between the city centre and its subcentres. This will be the result of the moderate compact spatial structure. These main links will be covered by metro and light rail connections. Trams and buses will operate on several links, though the importance of these modes will decrease somewhat through cuts in exploitation or operational subsidies. Short distances will often be covered by walking and cycling. In most cases conventional or electric and even new fuel cars will be used if the departure points or destinations are not in easy reach of the collective modes. Use of the private car will be made only slightly more expensive by increasing parking fees, but people will be able to easily afford these because of the anticipated economic growth.

In general, therefore, the same system as in the 1990s is expected to exist, which is a mix of many modes in which the car and its improved versions are dominant. The main change will be that the emissions of the car will largely diminish and that about 50 per cent of the cars will be electric or run on new fuels. Public transport will become more expensive for the users, because subsidies will be partly removed. Due to the compact structure of the cities these modes will maintain their market share, however.

The mobility growth in cities will be lower because of the moderate compact structure, but it will continue due to the growing population. The private car will account for about 60 per cent of the modal share (more or less equally distributed between electric/new fuels and conventional fuels), while the remaining 40 per cent will be covered by collective and slow modes.

10.4 THE DESIRED SCENARIO

Now we will turn to the presentation of the results of the *desired* future scenario as created by Dutch transport experts. When the scores on desired situations are analysed, it may be concluded that governmental measures, collective modes and new technologies should be introduced more widely than is the case in the expected scenario, while the use of the private car should be significantly reduced. On the other hand, the desired developments are less extreme than in either of the two reference scenarios.

The expert views given may again be positioned in the Spider model, which gives the desired result as presented in Figure 10.3.

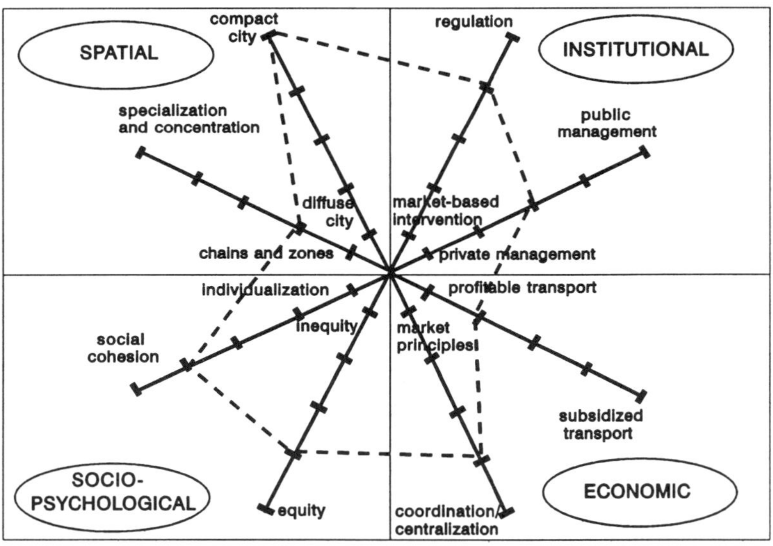

Figure 10.3. The Spider model: the desired scenario
Note: $n = 271$

It appears that on most axes the score is more in the direction of the outer limits of the Spider model, hence towards the regulatory scenario. However, there are two exceptions: in the European spatial organization a moderate diffuse spatial organization is desired, while the transport sector should become more profitable than in the expected case. Such interesting outcomes are the result of the menu-driven, non-packaging way in which the scenarios are constructed and thus reflect creative expert views on the future of transport. As discussed in

Section 7.4, these mixed views do not affect the basic consistency of the scenarios.

It appears that individual behaviour should change much more, while several dedicated policy measures and new technologies should be introduced to a much larger extent. The decision-making system becomes much more centralized and policies are mainly focused on collectivization of the transport system. In the next sections the distinct aspects and axes will be discussed, again using the Spider framework of analysis.

10.4.1 Spatial factors

The European spatial organization

In the desired scenario a policy is envisaged for Europe in which equal regional economic development is a main objective. As a result of this, activities and population should move (relatively) out of the core regions towards more rural and peripheral regions, and at the same time the main cities in these regions would become bigger and more important. Also the suburbanization trend should largely continue, resulting in a shift of population out of the bigger cities. This shift should mainly be oriented towards the smaller cities in the rural and peripheral regions.

Transport policy should focus on increasing the modal share of collective modes and reducing the use of cars. Therefore, most activities and people should become concentrated near nodes of collective transport modes in the bigger cities. These regional cities should become of strategic importance for the economy and society, since these are important economic, political and cultural centres. In general, they should have an HST terminal, and many offices and industries should be located there. The government should make it attractive for companies to move to these cities by offering grants, subsidies, etc. These may be justified by the argument that every region should be largely independent and self-supporting, while for environmental reasons the shift out of the highly populated and polluted core regions is justified.

Therefore, one may not speak of a classic hierarchical spatial organization, with a main city at the centre and smaller cities in its service area, although the number of inhabitants will differ per city. In this way a development according to the chains and zones model should occur, with more concentration in urban centres, thus being a moderate diffuse organization. The criss-cross transport links which are a result of this spatial organization should mainly be covered by an extensive network of collective modes, offering a high-quality service.

The urban spatial organization

The general policy should increase the modal share of collective modes and make car use less attractive. Therefore, cities should develop according to the compact city concept, in which a concentration of population and activities takes place around the main nodes of the collective transport system (HST, train and metro/light rail stations). In this way a spatial organization should emerge with compact city centres, and, in the bigger cities, several subcentres. Because the number of people per city should also be relatively low as a result of the European spatial development discussed above, living space in the cities may not become extremely expensive. This should also be the case because the average household size would become somewhat larger, again due to the reversal of the individualization trend. The demand for housing would be reduced a little by these trends, which would make housing cheaper. Most people, therefore, should be able to afford to live in these compact centres. In these centres it would make sense to continue to construct many subterranean and high-rise buildings to save space.

In the old quarters of the largest cities in the core regions especially, the lower income groups will live in old and not very well-maintained houses. Also the transport system providing access to these quarters may be of a rather low quality. Because of the political interest in equity, this group would be kept as small as possible, so that such poor living quarters can be kept to a minimum.

10.4.2 Institutional factors

Intervention by the government

In the view of experts it is generally desired that the influence of the government in many fields of society and the economy should be increased. The EU should develop as the most influential political authority, which should introduce many compulsory standards and targets, while a regional support programme should also be one of the main aims of EU policy. The national authorities should become relatively unimportant. The regional/urban governments on the other hand need to gain much more influence, because of the subsidiarity principle, which implies that every decision should be taken at the lowest policy level possible. Many responsibilities should therefore be handed over to this level.

A strict environmental policy should be set up at the European level, to reduce free rider behaviour by the Member States or the regions. Many standards and tariffs should be introduced, for example, in the transport market. One of them should be a compulsory price increase of conventional fuels; this would make the new fuels and electric cars so attractive that conventional cars would largely vanish from the market. In society these measures should be accepted, because they are necessary for environmental reasons and for improving the well-being of people. This acceptance should also be stimulated by the fact that the individualization trend is reversed implying there is more social awareness.

Also, to increase equity in society, a European social security system should be introduced, as well as other (e.g. regional and agricultural) support programmes. To reinforce these measures a strategy should be introduced to protect the economy from competition by other trade blocks in the world.

Organization of the transport market

As mentioned above, transport should be the responsibility of the EU as far as the main links and the railways are concerned (e.g. trans-European Networks), while regional and urban authorities should be responsible for the urban and regional public transport companies.

It is seen as desirable for car use to become much more unattractive as a result of increases in (conventional) fuel prices, parking fees, and a road pricing system should be introduced on many links. Also the number of parking places should largely be reduced in the cities, again to make car use less attractive. This space is also needed for the construction of compact city centres.

At the same time, large-scale investments by the EU and the regional authorities should take place in a collective transport system. This network should cover all main transport links. The railway companies should become much more independent of government influence. This may be possible, since the government may accept the responsibility for the provision and maintenance of the infrastructure, which would reduce the costs for the operators. Moreover, the exploitation of rail infrastructure should be made more independent of the government. This may be possible, because the losses would be largely reduced due to the higher demand per link, caused by the more expensive and therefore reduced, use of the car.

The collective modes in the urban transport system should ideally be the responsibility of regional authorities. Therefore, these public transit companies could also become more independent from the government. As a result, the modal share should change in favour of collective modes, and the use of the car should be greatly reduced.

10.4.3 Economic factors

Development of the European economy

As discussed above, in the desired scenario much more coordination of the economy should be performed by the EU. However, this would have negative impacts on economic growth, but may be accepted as necessary to save the environment and to achieve more regional and social equity. Therefore, the European position in the world economy would decline relative to the United States, Japan or the Pacific region. The free market should not necessarily be accepted as the best way to achieve the policy goals, and therefore strict regulations should be introduced in many cases.

Because of regional support mechanisms and a policy dedicated to achieve equal regional development all over Europe, a shift out of the core regions towards the other regions should be stimulated. This trend should be reinforced by spatial policy and public support to encourage people or firms to move out of these regions to urban centres, by means of subsidies and taxes. Therefore, the core regions would relatively decline, while the rural and peripheral regions would enjoy a positive economic development. As a result, the disparities between regions would be reduced.

Profitability of the transport market
It is desirable for the government to intervene strongly in the transport market and oblige public transport operators, notably railway companies, to cover all transport links, and therefore the network and its capacity should be greatly increased. At the same time fares should be kept at a low level, to make these modes more attractive for everyone. Despite this, the profitability of railway companies should increase significantly. This should not only occur in long-distance transport where the collective modes have traditionally had the best competitive position, but to a lesser extent also in the regional transport system. This should be achieved by increasing the efficiency of the railways, a concentration of population in the urban centres and large-scale investments in new infrastructure (by the government). Also, the transport links should become more dense because of these trends.

The profitability of the railways should also be increased by actively discouraging car use, which will improve the competitive position of these modes. For financing road infrastructure, for example, road-pricing systems should be introduced everywhere. Nevertheless, because of the reduction of car traffic, insufficient revenues may be raised by these systems to privatize the infrastructure. Therefore, the government should still own and plan the infrastructure.

10.4.4 Socio-psychological factors

Individualization
It should be increasingly accepted that individualization and its consequences cannot be allowed to continue, since this implies too many negative impacts for the environment and the welfare of individuals. Therefore the individualization trend should be stopped and even reversed to some extent. As a result, the average number of people per household may grow somewhat, which should reinforce the compact structure of cities, thus making housing more expensive.

This reversal should have positive impacts on the acceptance of environmental measures, which should be introduced on a large scale. In the transport market, for example, it should be accepted that the use of collective modes should be stimulated and car use should be reduced. Evasion, however, is a serious problem, because free rider behaviour is attractive for all individuals.

Equity in society
In the normative view of experts, equity is desired to become an important goal of the EU and in regional policy. Therefore, funds should be raised to achieve more equal regional development and to maintain a quite extensive social security system, although this may have negative consequences for economic growth.

Three economic groupings may be distinguished. The highest income earners (about 5 to 15 per cent of the population) can afford to live outside the compact city centres in 'green' suburbs and eventually outside the cities altogether. They own a car - run on new fuels and electricity, although they should use it sparingly, preferably only to reach the nodes of the collective modes, because their low density living areas are not within easy reach of collective modes. Car trips should become very expensive because of high fuel prices and road pricing systems; so this is the only group which can afford to drive a car without any problem.

About 75 to 85 per cent of the population will belong to the middle income earners. This group will mainly live in the compact city centres, and prefers not to own a private car, because of the high costs and the scarce space in these centres. Even so, they need to live near the nodes of collective modes. When they need a car, for example, to reach recreational areas or nature areas, they may rent one at convenient and easily accessible places.

Finally, the lowest income earners comprise of about 5 to 15 per cent of the population, who will mostly live just outside the compact centres of the cities, in quarters which may only be reached by low-quality modes like bus or tram. These people may be entirely dependent on collective modes for their travel needs, which may seriously limit their mobility possibilities.

10.4.5 The transport system in the desired scenario

Long-distance transport
In general, it is desired that a transport system exists with many criss-cross transport links between the larger cities. These links should be covered by an extensive network of HST and fast train links.

Private car use should largely be decreased for long distances. Only on links where no efficient rail infrastructure is available should cars, run on new fuels, be used. On each highway a road pricing system should be introduced, which should considerably increase the costs of car use and also stimulate car pooling to share these costs. Very long distances should be covered by LH_2 planes, which would replace the services of conventional planes almost entirely. These should offer a service only where there is no high-quality HST link available (especially intercontinental), because otherwise they are not competitive.

Subterranean transport systems for conventional and new modes (shuttles) and the Maglev train are not be introduced on a large scale, since these are too expensive and not compatible with existing modes.

Also the main regional links should be covered by trains, or for the smaller distances by metro or light rail systems. For transport to places where no - or hardly any - collective modes are available, private cars should be used. These cars are mainly used by the high-income groups, who live in low-density areas where no collective transport is available, but these classes can afford these cars. They should use them mainly to reach main nodes of public transport, because this is more attractive than cars for many trips.

It is hoped that the mobility growth over the longer distances should stop at least temporarily, when the private car is made more unattractive. However, when rail infrastructure is constructed, efficient railways should offer a high-quality service. As a result, mobility may start growing again. This growth may then be relatively high, because the main economic activities would be spread all over Europe, which increases travelling distances and needs.

The modal split should largely change in favour of the collective modes, which may account for some 75 per cent of travel demand. Especially in regional traffic the private car (driving on new fuels) should have a market share of about 20 per cent, while the LH_2 plane has a share of about 5 per cent.

Urban transport

In the desired scenario, the cities in the year 2030 are developing compact spatial structures with several subcentres, in which a radial transport system forms the basic network. These rapid transit high-density links should be covered by metro and light rail systems. These transit systems should form the backbone of the urban transport system and should connect activity centres with each other and with the nodes of HST and other trains.

Over very short distances, walking and cycling should be used as the main modes, since the car should not be used here, while collective modes are not efficient over short distances. Sometimes, however, people movers could also offer a convenient service on short distances with a high density of demand.

At the nodes of the metro and light rail systems, trams and buses should offer a service to the less populated areas where mainly the lower-income group lives. At the borders of urban areas parking places should be constructed for private cars, for use primarily by the wealthier people. These cars are run on electricity or on new fuels, since conventional fuels are too expensive. In this spatial configuration, they should not drive into the cities, where space is scarce and driving very expensive. In fact, the use of cars should be deliberately made very expensive; to achieve this sometimes a road pricing system should be introduced, while also parking fees should rise.

Mobility growth is desired to be negative when car use diminishes. Later, this growth rate should increase, however, because of economic and population growth. About 90 per cent of urban transport should consist of collective and slow modes, and the remaining 10 per cent cars, which are almost all run on new fuels and electricity.

10.5 REALITY AND WISH

The scenarios described above have generated a wealth of information on the driving forces, features, opportunities and bottlenecks of alternative futures and their transportation consequences. When the *expected* scenario is investigated, it appears that the trends and underlying factors which are discussed in Part A will not dramatically change. For example, no large-scale changes are expected for individual citizens; they are not expected to drastically change their behaviour in order to reduce the externalities caused by transport. The same holds for the government, whose policy is expected to be inadequate; only spatial policy may be rather successfully implemented. Neither will new transport modes and technologies, like new fuels (e.g. hydrogen) or collective modes be widely introduced. On the whole, therefore, the system and society may more or less be the same as today. Also mobility growth will largely continue, although at a lower pace, because of the reasons discussed in Section 3.3.2. It is doubtful whether large-scale reductions of externalities may be achieved in this scenario, because if sustainability criteria are to be achieved, a much greater improvement of current modes (car and plane) would have to be achieved than is anticipated at present.

In the *desired* scenario, there is a clear choice for a collective transport system. Therefore, large-scale changes are necessary in the behaviour of individuals, as well as in the institutional and economic environment. At the same time the government should invest heavily in infrastructure and take measures to reduce car use substantially, though at the same time mobility growth should slow down significantly. It is evident that it would be very hard for the government and society to let this scenario become reality. In this respect, it is interesting to note that more than 80 per cent of the respondents answer ‘may be’ or ‘no’ to the question about whether the transport system will be sustainable in the year 2030. Clearly, there is a wide gap between reality and wish, and the experts fully expect that as things stand the transport system in the year 2030 is likely to be unsustainable, even though it would be desirable to implement drastic measures to achieve a sustainable one.

It would, of course, be interesting to know how the interviewed experts judge the final scenarios which have been constructed on the basis of individual views and expectations expressed by themselves. Are these scenarios *ex post* regarded as meaningful or plausible? This will be the subject matter of the next section.

10.6 INTRODUCTION TO SCENARIO ASSESSMENTS

In the previous sections of this chapter an expected and a desired scenario for sustainable transport systems have been constructed based on Dutch expert opinions. The design of these scenarios was partly carried out on the basis an intuitive interpretative analysis of the results obtained from the survey questionnaire (see Section 10.2). Consequently, the scenarios may certainly contain some value judgement of the designers.

This makes it interesting to analyse what the respondents in this survey think of the scenarios created from their views and how far the scenarios conform to their expectations and wishes. In order to receive a second expert opinion, the possibility was offered to the respondents-experts of the first-round questionnaire to ask for the results of the study. Therefore, we had the opportunity to send a summary of the scenarios to the respondents who asked for the results (Nijkamp *et al.*, 1997), together with a second short survey questionnaire with a limited number (six) of questions on the plausibility, realism and sustainability of both the expected and desired scenario. The response rate of this survey was 22 per cent ($n = 43$). The results of this are described in Section 10.7.

A second feature of the scenarios is that they are constructed only on the basis of the opinions of Dutch experts. As a result, there may be a certain bias in the answers, because of the specific Dutch situation regarding transport, land use and the environment. This makes it interesting to find out how experts from other countries would judge these scenarios in order gain additional insights from some sort of international contrast analysis. Two re-assessment possibilities were used to acquire some new views on the overall value and validity of these scenarios. First, the results have been discussed at three successive international workshops attended by European transport experts. And second, the same short second-round questionnaire was handed out to the participants at the Euro-conference on 'European Transport Networks and Policies' (Espinho, April 17 to 22, 1995) at which a summary of the scenarios and their results were presented. Twenty-five experts responded to this questionnaire. In Section 10.8 we will discuss the views of these European transport experts. In the last section we will draw some general conclusions on these scenario experiments.

10.7 A SELF-ASSESSMENT OF THE EXPERT-BASED SCENARIOS

In the second-round questionnaire, six specific questions were posed to the Dutch transport experts. In addition, there was also the possibility to give additional comments on each question. Regarding the *expected* scenario it was asked whether the scenario conformed to their own expectation (measured on a 5-point scale from 'not' until 'fully'; the same scale was used for all 'non-open' questions as well). The second question was whether this scenario would lead to the achievement of general sustainability criteria.

For the *desired* scenario the validation question was split into two questions in order to make a distinction between the main characteristics of the scenario:

- Do you agree with more regulation?
- Do you agree with a collectivization of transport?

Next, as with the expected scenario, it was asked whether this desired scenario is regarded as sufficient for achieving general sustainability criteria. Finally, an open question was added on the perceived bottlenecks, should the desired scenario become reality. We will now successively discuss these second-round results for both the expected and the desired scenario.

10.7.1 Results of the self-assessment of the expected scenario

The results on the questions related to the self-assessment by the Dutch transport experts regarding their expected scenario are presented in Figure 10.4.

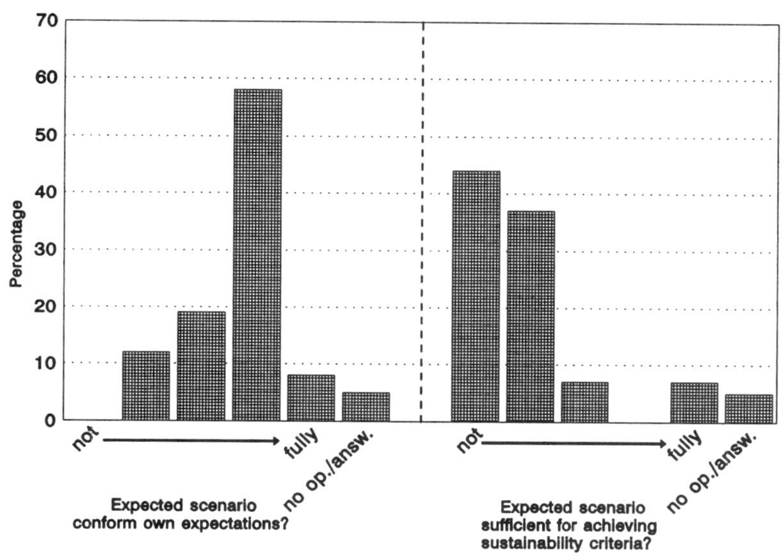

Figure 10.4. Opinions of Dutch experts on the expected scenario
Note: $n = 43$

About 60 per cent of the respondents give the value '4' - on a 5-point scale running from 'not' until 'fully' - to the expected scenario. Only 12 per cent give the value '2', while nobody totally disagrees with the scenario. Also the comments made are rather positive. Mostly the scenario is regarded as an extrapolation of current trends, although some think the expected scenario is too pessimistic. So it may be concluded that the scenario is a good presentation of the expected future as seen by the Dutch experts.

On the question concerning whether the expected scenario will lead to an achievement of general sustainability criteria, an overwhelming majority of 81 per cent expresses a negative opinion ('no' or 'hardly'), while only 7 per cent answers completely affirmatively. In the comments it is said that especially emissions of harmful gases - at both the local and global level - will not be reduced unless there are drastic improvements in current technologies; other externalities may be reduced to a greater extent.

It may thus be concluded that most respondents do not expect that the transport system will to a large extent become sustainable. This is obviously a rather pessimistic view of the potential outcome of current transport and spatial policy, but the results are largely consistent with the expected results presented in Section 3.9.2.

The aggregate results on the opinions of Dutch experts regarding the desired scenario which was constructed from their *a priori* expressed views are presented in Figure 10.5. Besides these three questions, it was also asked which bottlenecks are to be expected for this scenario, should it become reality.

10.7.2 Results of the self-assessment of the desired scenario

About two thirds of the respondents appear to support the choice in the desired scenario for the introduction of more stringent regulatory measures than in the expected case, although it is doubted whether this is politically feasible. Some of the respondents who answer 'not' argue that only market-based measures (internalizing external costs in the market price) will be successful. Also more than 50 per cent of the respondents support the idea of a more collective transport system. Problems are mainly seen in the behavioural and psychological fields. Others mention that this strategy will only be successful for the longer distances. It may be concluded, therefore, that the main substance of the desired scenario presented in the previous chapter more or less conforms to the idea of the majority of the respondents.

Another interesting question is whether this scenario will lead to a better achievement of general sustainability criteria. About 45 per cent of the respondents gives the value '4' or '5' for this question, which means that most respondents think a scenario like the desired one may lead to a more or less sustainable transport system. Some think, however, that the living conditions will worsen if this scenario becomes reality. Some experts also think that the mobility growth should be lowered much more.

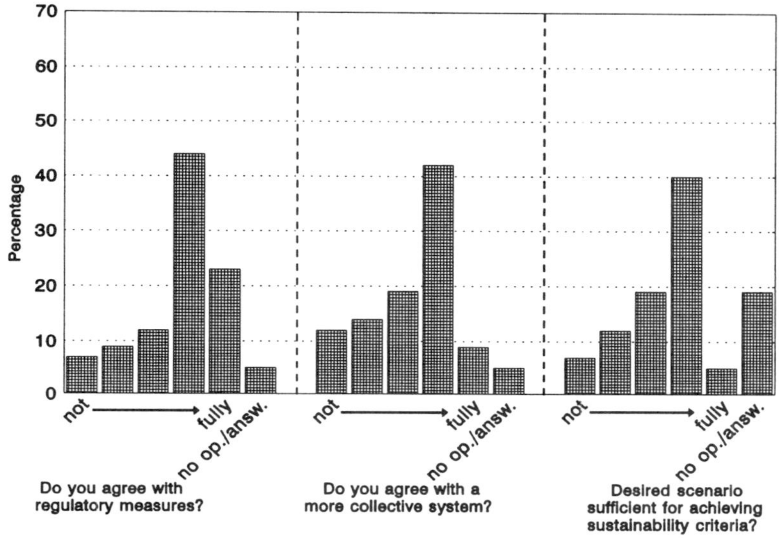

Figure 10.5. Opinions of Dutch experts on the desired scenario
Note: $n = 43$

Bottlenecks are seen in the following fields:

- the scenario is not politically feasible;
- car manufacturers will oppose the scenario;
- such a policy will not be accepted by the population;
- financial problems because of the large-scale investments needed;
- current trends are difficult to curb (individualization, deregulation, etc.);
- the EU is not expected to become a very influential organization, so the desired international cooperation will not take place to any high degree.

So it may in general be concluded that both the expected and desired scenario are well in tune with the opinions of the large majority of the respondents/experts. This also holds when the general comments are analysed. The expected scenario is not thought to result in a sustainable transport system, whereas the desired scenario may largely result in the achievement of those criteria necessary for sustainability, at least according to the second-round views of the experts. The actual implementation of the policies needed to achieve this scenario will, however, be difficult.

Next, it is fascinating to see whether transport experts in other European countries appreciate these scenarios in the same way as the Dutch experts. This will be analysed in the next section.

10.8 A REASSESSMENT OF THE EXPERT-BASED SCENARIOS BY EUROPEAN EXPERTS

After the self-assessment it was decided to organize a reassessment by European transport researchers. The same survey questionnaire that was given to the Dutch experts was also handed out to a representative group of several European experts, in order to analyse the extent to which their opinion would differ from the average Dutch view. Of course, the number of respondents (25) is too small to draw solid statistical conclusions, but the results may at least provide some interesting information on the opinions of European experts regarding the views of their Dutch colleagues. In the same way as in the previous section, we will first discuss the opinions on the expected scenario and then on the desired scenario.

10.8.1 Results of the reassessment of the expected scenario by European experts

The aggregate evaluation results offered by European experts on the expected scenario are presented in Figure 10.6.

It turns out that more than 50 per cent of the experts support the expected scenario by giving a score value of at least '4'. Though a large minority has a neutral opinion, only a very small share disagrees with the expected scenario.

Several respondents mention in their comments that they do not expect the moderate compact city to come into existence, but that they expect more a development towards the diffuse city. In general, however, it may be concluded that the main trends reflected in the expected scenario are also supported by the European experts, but to a slightly lesser extent than by their Dutch counterparts.

It is noteworthy that an even larger share of the European experts is of the opinion that this scenario is not sufficient for achieving sustainability criteria. Thus in this respect also the average European view appears to coincide largely with the opinion of the Dutch experts. In the next subsection we will turn to the European opinions on the desired scenario.

10.8.2 Results of the reassessment of European experts on the desired scenario

The final stage of the reassessment analysis again used the same questions as for the Dutch experts. The results are presented in Figure 10.7 and offer some interesting information.

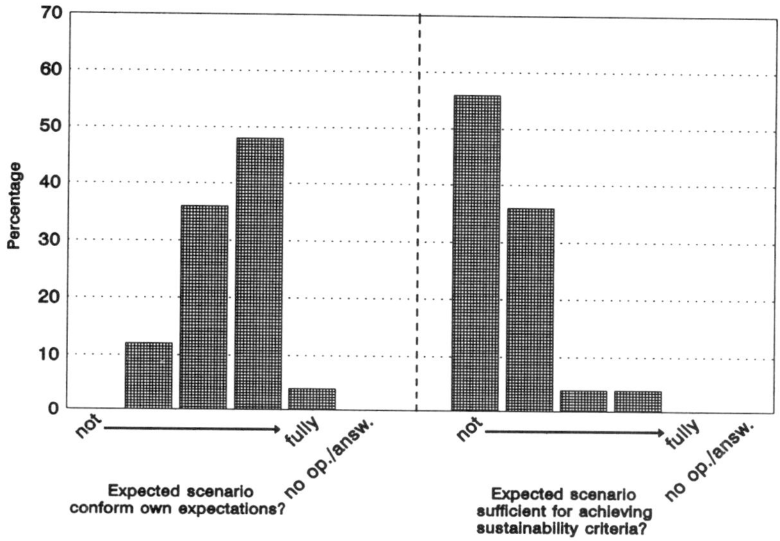

Figure 10.6. Opinions of European experts on the expected scenario
Note: $n = 25$

It is first of all noteworthy that, although a very large minority gives regulatory measures scores of only '1' to '3', the actual majority strongly supports the choice for more regulatory measures, although several respondents prefer market-based measures such as road pricing. The same response holds for the choice for a collective transport system.

Next, it also appears that the European experts are less convinced when confronted with the question of whether this scenario will result in the fulfilment of general sustainability criteria. In fact, nobody thinks this will be fully the case. Furthermore, about one third of the respondents give the neutral value '3', while about one third think this is not the case. So it may be concluded that the European experts are even more pessimistic about the achievement of sustainability criteria than their Dutch counterparts. In summary, the attainment of a sustainable development appears to be fraught with many difficulties, even in an ideal-type of scenario.

10.9 CONCLUSIONS

It may be concluded that by and large both the expected and desired scenario are affirmatively valued by both Dutch and European experts. Large majorities appear to support the validity of these scenarios with a value of at least '4' on

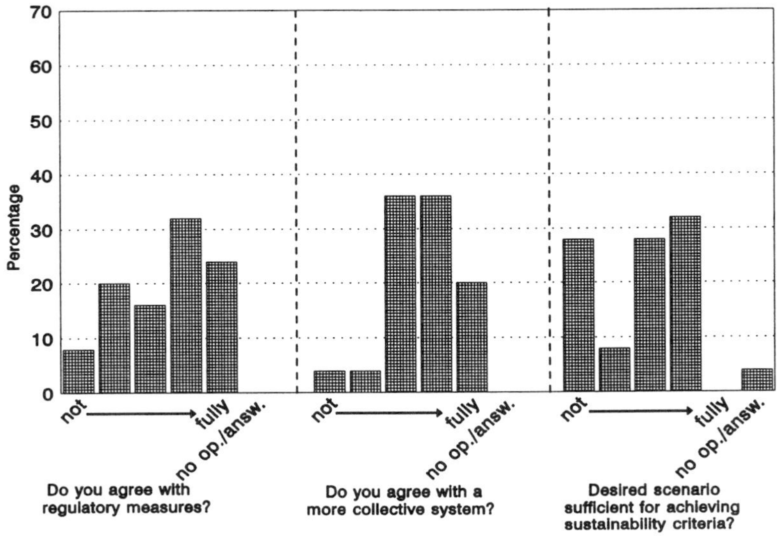

Figure 10.7. Opinions of European experts on the desired scenario
Note: $n = 25$

a 1 to 5 scale. However, according to both Dutch and European experts, the *expected* scenario will most probably not lead to the achievement of sustainability criteria. However, there is no such convergence of Dutch and European opinion about whether the *desired* scenario is sustainable. The Dutch experts think in general that the *desired* scenario may be sufficient for this objective, while the European experts are less convinced on this point.

Thus it turns out that experts may well agree on the desirability of certain new spatial and transport options, but they may somewhat disagree on the foreseen sustainability implications of such alternatives. This is, of course, to some extent an empirical question which requires an assessment of the environmental consequences of each of these options. Therefore in Chapter 12 we will offer an answer to the question and analyse how far the scenarios - including the reference scenarios - will lead to reductions of the CO_2 emissions caused by the transport sector. In this way, we will be able to resolve the extent to which these scenarios may be sufficient for achieving this particular sustainability criterion.

Before moving to the environmental assessment of the scenarios, one important question has to be addressed, viz. the diversity of opinion about both the expected and the desired scenario. If there is a large discrepancy among expert views, it makes less sense to compose aggregate scenarios, as then interesting contrasts - which may be more innovative - are left out of consideration. Evidence for this assumption can be found in the broad scenario study by Masser *et al.* (1992). Therefore, in the next chapter, we will focus on a segment analysis of the Dutch transportation experts' views in order to investigate whether alternative, interesting, minority or contrasting views can be identified.

11 A SEGMENT ANALYSIS OF EXPERT GROUPS FOR TRANSPORT SCENARIO DESIGN

11.1 A NEED FOR ALTERNATIVE VIEWS

Scenarios should not only view a common sense future, but also focus on interesting minority viewpoints including contrast-images of the future. This means that in both our expected and desired scenario construction sufficient attention has to be given to alternative views. Even though the distribution of the expert opinions was rather uniform, it is an interesting research endeavour to try to identify homogeneous subgroups with rather similar viewpoints.

One of the advantages of applying the expert-opinion, non-packaging approach is that, in principle, scenarios can easily be constructed on the basis of minority opinions from experts (see Chapter 7). A systematic way of investigating such interesting options is to undertake a segmentation of the respondents which may be expected to lead to interesting new and deviating patterns of opinion. Therefore, we will now investigate to what extent differences in disciplinary background, job, gender or age of respondents may cause different views on the future of transport. In this way we are able to construct more segment-specific Spider scenarios for the future of transport. In the present chapter an analysis will be made of the different scenarios for four distinct categories of respondents. Next, it will be shown how different scenarios can be constructed by using the opinions of these subsegments.

11.2 DESCRIPTION OF SEGMENTS

In this section we will describe in more detail the four subsegments of respondents, including some tentative aggregate results.

Educational differences

A principal distinction in our sample can be made between three main disciplinary categories: *traffic engineers*, *(transport) economists* and *planners*. From the answers to the survey questionnaire, we find that planners in general *expect* more social cohesion in society, while economists expect (and desire) less behavioural change and more measures for reducing car use (like road pricing, higher parking tariffs, etc.). In general, however, the differences turn out to be small compared to the average pattern. The same holds for the scores on the distinct transport modes. Therefore, we conclude that disciplinary backgrounds apparently do not lead to drastic differences in the general expert scenarios. Consequently, a disciplinary analysis will not lead to new contrast scenarios.

Professional differences

In the analysis per profession a distinction has been made between five main professional categories: *municipal officers* (including those working in provincial authorities), *university staff*, *consultants*, *ministry officials* and *experts working for public transport companies* (railways, regional and urban transport).

It appears - perhaps not very striking - that the municipal respondents *desire* that transport decisions will be taken at the regional level. In the *expected* spatial organization it is interesting to note that ministry respondents expect a moderate diffuse - instead of a moderate compact - spatial organization. The university staff *expect* that mobility growth of cars will be more or less equal to that of public transport in the urban transport case.

When the behavioural and policy options are analysed, it appears that the ministry officials *desire* - and *expect* - the least reduction in car use and also the least stimulation of collective modes. The opposite opinion is found for public transport company and municipal officers, who also *desire* a more active stimulation of collective transport mode use.

As a result, it may be concluded that some of the axes of the Spiders may differ somewhat per employer category, but that obviously no big differences are found. The main difference is that the transport system should be more collective according to public transport company experts and to a lesser extent according to municipal officers.

Gender differences

The analysis of gender differences reveals interesting features with respect to the Dutch transport experts. More than 90 per cent of the respondents appear to be male and only 7 per cent female (3 per cent unknown), thus it is clear that the male answers are about the same as the average ones.

The female respondents, however, differ clearly in several ways from the average scores. First, they *expect* the main decisions to be taken at the European level, while they *desire* these to be taken at the national level. In the spatial organization at the European level a strong spatial concentration is expected. This gives a slightly different expected scenario, as presented in Figure 11.1.

In general, the female scores for the *expected* behavioural and policy options to reduce car use are lower, while in the view of females the use of collective modes will be stimulated less than in the average (male-dominated) scenarios. The *desired* scores, however, are about the same as the average ones. In the transport system, collective and new modes get high scores, too, so that a more collective-oriented system may emerge, as in the expected scenario of all respondents.

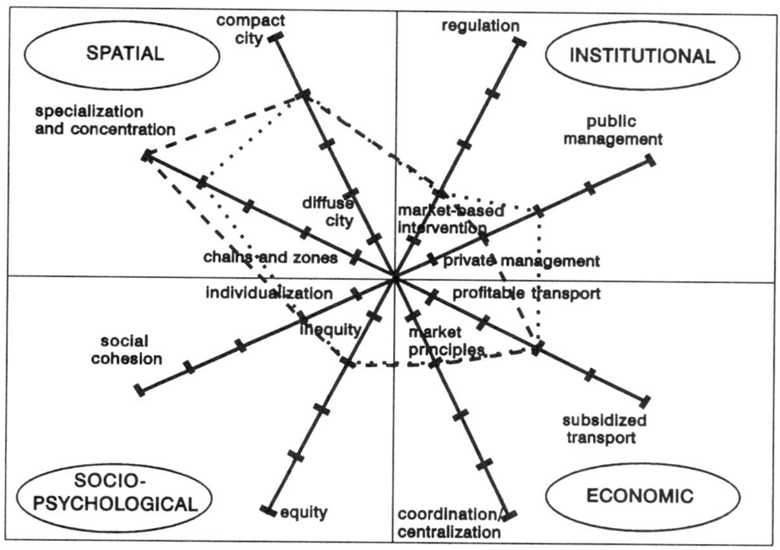

Figure 11.1. Expected scenario of female respondents
Notes: 1) $n = 15$
2) .... male (= average); -------- female

Age differences
Finally, we will also look at differences in responses according to age category. A distinction was made here between four age categories: younger than 30, 30 to 40, 41 to 50 and older than 50 years. When the respondents' answers are analysed, it appears that the youngest category has a slightly atypical view on the future of transport. In particular, their expected scores about future behaviour of users suggest that car use will be reduced much less. This holds for the scores for driving fewer kilometres by car, fewer car days, etc. These young researchers also give lower scores to a rise in fuel prices, parking tariffs, etc. The oldest category of experts on the other hand has in some cases a rather high preference for a somewhat greater restriction of car use. Thus, it seems that the views of the younger generation are slightly more 'conservative' than those of the older age categories.

For the remainder, however, the answers given on developments per age category do not differ significantly; the same holds for the scores on the modes in the future transport system. So it may be concluded that the differences per age category do not lead to entirely different expert-based scenarios.

11.3 SCENARIOS PER CROSS-CLASSIFIED SUBCATEGORY OF RESPONDENTS

Although no large differences have been found in the four major distinct categories - and thus hardly any discriminating scenarios could be identified for specific groups, it is interesting to distinguish cross-classified subcategories for the respondents, in order to analyse whether different scenarios can be constructed in this way. Of course, this exercise is constrained by the number of cells - in particular empty cells - in a contingency table analysis. We will take as a minimum level per subcategory a total of 10 respondents. We will only present Spiders and scenarios descriptions of those subgroups for which we find different scores on at least two axes. The opinions of other subgroups are only assessed in table form.

11.3.1 Professional and disciplinary characteristics

In our analysis of cross-classified subsegments of respondents we will mainly focus on professional and disciplinary differences - and to some extent also on age differences, - as here some interesting contrasts for designing new scenarios can be found. From a two-dimensional contingency table, we will now discuss some interesting pairwise groups in the next three subsections.

Traffic engineers working at a ministry

The traffic engineers working at a ministry appear to have a different view from the average on the development of the future spatial organization. They *expect* a moderate diffuse organization instead of a moderate compact one, whereas a mixed development is *desired* (more compact than in the average desired scenario). At the urban scale level a mixed development is *expected*, which is less compact than the average view. Urban mobility growth is *desired* to come to a standstill. Decentralization is both *expected* and *desired*, because decisions should be taken at the regional level. For the remainder the scores are about the same as in the general expert scenarios. For the behavioural and policy options, this subcategory gives relatively low scores to the option of driving fewer kilometres and days per car. On the other hand, low scores are also found for the use of collective modes and road-pricing systems. Moreover, the scores for the policy measures indicate that car use is not expected to be limited very much. The resulting *expected* and *desired* scenario may then be drawn as presented in Figures 11.2 and 11.3.

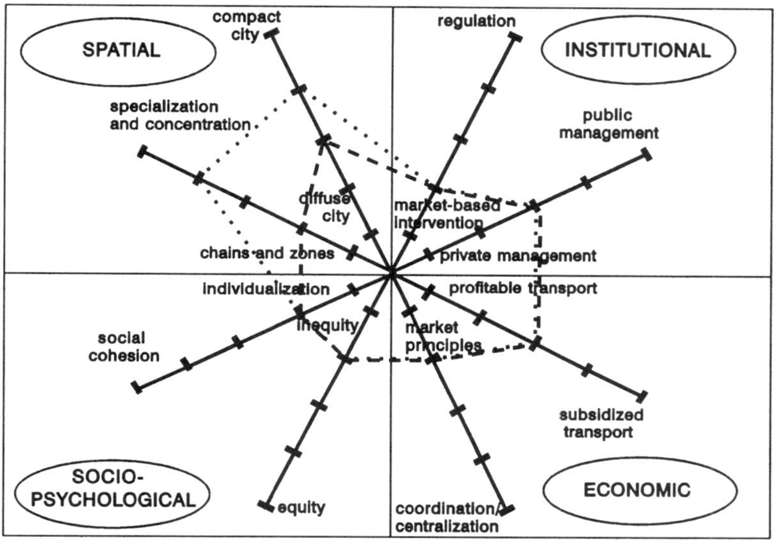

Figure 11.2. Expected scenario of traffic engineers working at a ministry
Notes: 1) $n = 18$
2) .... average; -------- traffic engineers, ministry

The scores of this subcategory for the various modes of the transport system do not differ significantly from the average scores, so that they expect and desire about the same system as in the average scenarios.

Traffic engineers working at a university
The expected spatial organization differs for this category from the average *expectation*: at the European level the expected spatial organization is entirely diffuse, while the urban development is mixed. The resulting *expected* scenario is presented in Figure 11.4. In the *desired* answers it is striking that the individualization trend should be reversed.

The scores for the various modes and measures are about the same as the average pattern, despite the specific different scenario views on how this transport system should be achieved.

The main differences between the average expected and desired scenario and the scenarios of the subsegments of professional and disciplinary backgrounds are summarized in Table 11.1.

The conclusion is that this cross-classified analysis brings to light some interesting findings, but these are certainly not very significant nor do they lead to sharply contrasting scenarios. In the next subsection we will also investigate the potential impact of age.

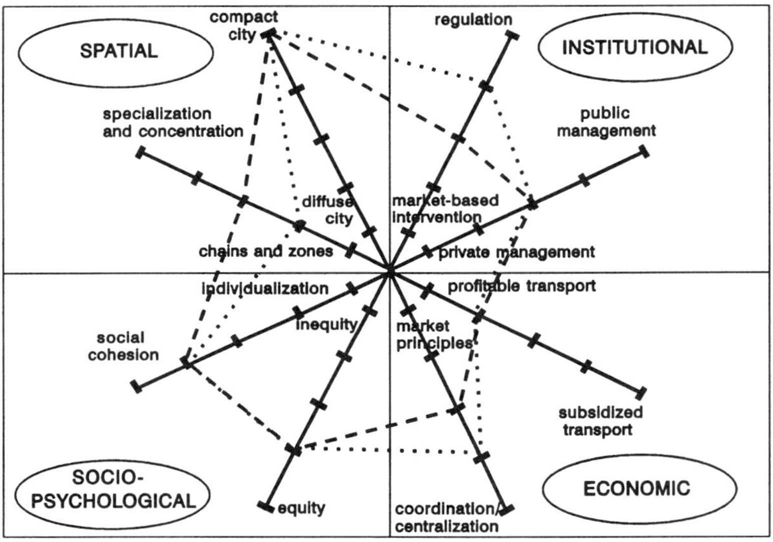

Figure 11.3. Desired scenario of traffic engineers working at a ministry

Notes: 1) $n = 18$

2) . . . . average; -------- traffic engineers, ministry

11.3.2 Disciplinary and age characteristics

Economists < 30 years

For this subcategory we find some interesting features. Expected behavioural changes which reduce car use are given low scores from the young economists, while the desired changes are about the same as the average scores.

For the policy options, both the expected and desired scores are low. Less intervention of the government is desired in the transport system, while the main decisions are to be taken at the European level. It is striking that the expected and desired profitability of the railways is regarded as low. The resulting scenarios are presented in Figures 11.5 and 11.6.

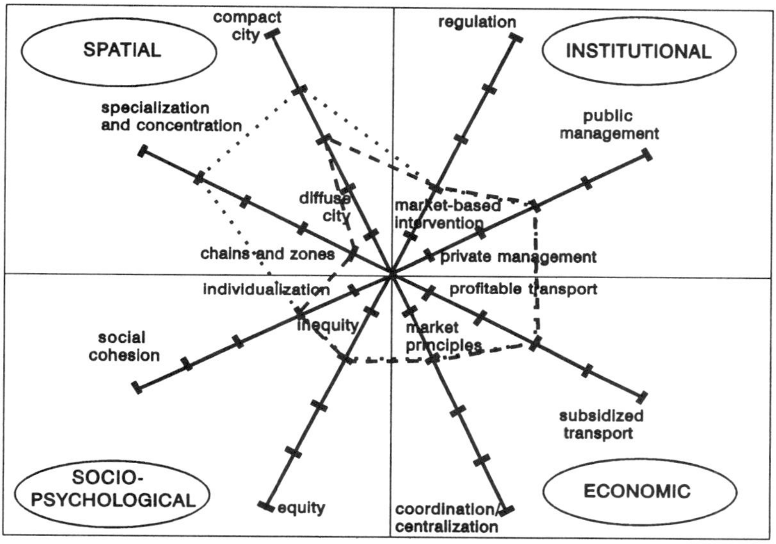

Figure 11.4. Expected scenario of traffic engineers working at a university
Notes: 1) $n = 14$

2) .... average; -------- traffic engineers, university

The resulting transport system is more individually-oriented, especially in the expected scenario: conventional as well as new fuel cars get relatively high scores, while road pricing especially is valued low. Also striking are the relatively high scores for Maglev trains and subterranean shuttles.

Planners 30-40 years

An interesting category is made up by planners in the age cohort of 30 to 40 years. It is unusual that this subcategory thinks economic growth is possible without mobility growth. A large impact is also expected from fuel price increases on car use. The expected spatial organization at a European level is mixed, though at the urban level moderately diffuse. For the remainder the answers given are about the same as the average ones; the resulting expected scenario is presented in Figure 11.7.

Table 11.1. Main scenario differences of subsegments of professional and disciplinary backgrounds

	Traffic engineers	Economists
Ministry	**Expected:** * Moderate diffuse European spatial organization * Mixed urban spatial organization * Decentralization * More car use **Desired:** * Mixed European spatial organization * Less urban mobility * Decentralization	
Municipality	**Desired:** * National level important * Car use more restricted	
Consultancy	**Desired:** * National level important * Mixed European spatial organization	**Expected and desired:** * Less restriction of car use
University	**Expected:** * Diffuse European spatial organization **Desired:** * Individualization reversed	
Public transport company	**Desired:** * Modal shift towards collective modes	No large differences

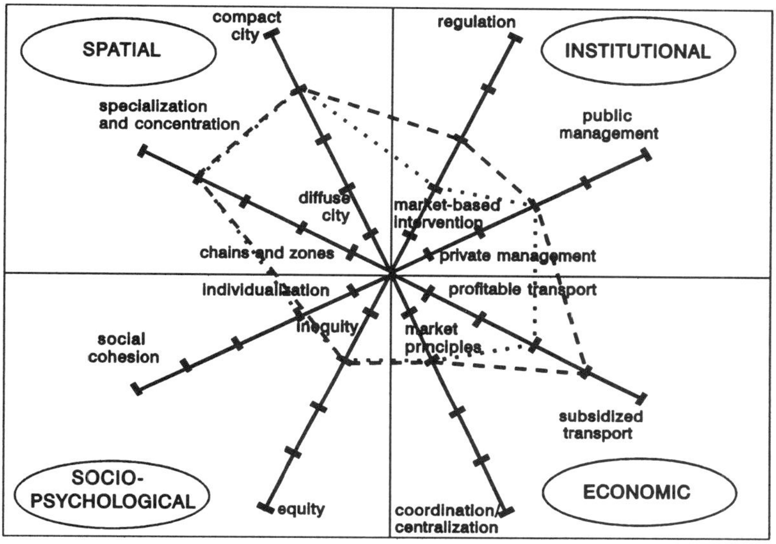

Figure 11.5. Expected scenario of economists, age cohort < 30 years
Notes: 1) $n = 10$

2) .... average; -------- economists, < 30 years

The resulting transport system corresponds approximately to the average scores and answers, with the exception of the desired urban transport system, which places somewhat more emphasis on collective modes.

The main differences of the subsegments with respect to disciplinary and age backgrounds are summarized in Table 11.2. As was the case in the previous subsection, it can be concluded that the differences are relatively minor. Next we turn to subsegments of respondents with respect to professional and age backgrounds.

11.3.3 Professional and age characteristics

Ministry, 41-50 years
It is noteworthy that this subcategory of officials assigns low scores to expected measures which make car use less attractive, although they desire more use of collective modes. They also give low expected scores to most options which increase the profitability of the railways.

The main decisions in the field of transport are expected to be taken at the European level, which indicates an increasing centralization. It is surprising that almost no impact on car use is expected from increases in the fuel price.

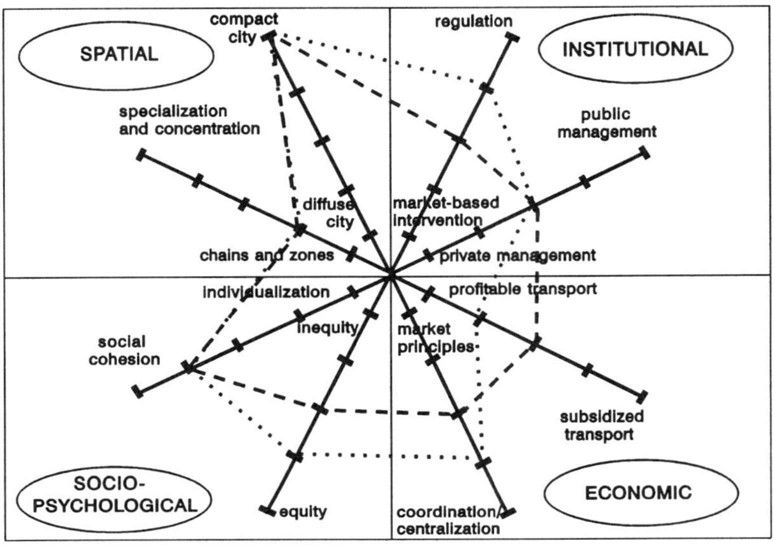

Figure 11.6. Desired scenario of economists, age cohort < 30 years

Notes: 1) $n = 10$ 2) .... average; -------- economists, < 30 years

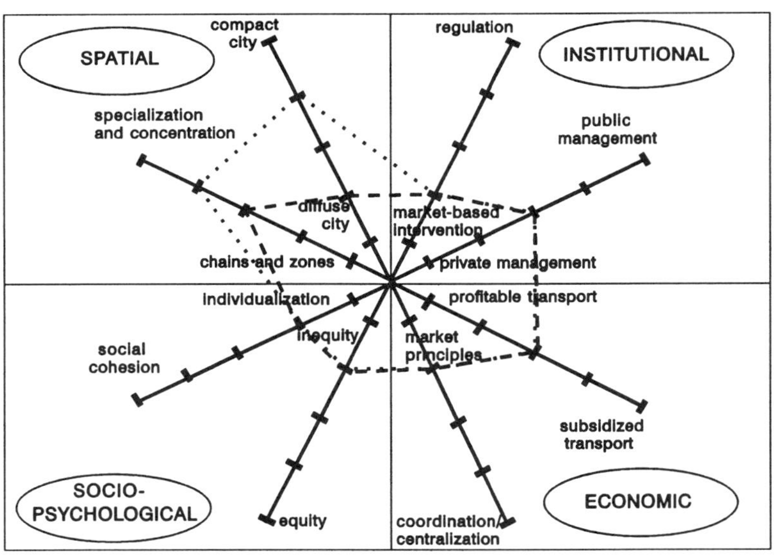

Figure 11.7. Expected scenario of planners, age cohort 30-40 years

Notes: 1) $n = 20$ 2) .... average; -------- planners, 30-40 years

Table 11.2. Main scenario differences of subsegments of disciplinary and age backgrounds

Age	Traffic engineers	Economists	Planners
< 30	**Expected:** * Much car use **Desired:** * Less centraliz. * Much car use	**Expected:** * More car use * Profitability railways low **Desired:** * More car use * Less intervention * Profitability railways low	
30-40	**Expected:** * Mixed European spatial organiz. **Desired:** * Restricted car use	**Expected:** * Less use of collective modes * Mixed European spatial organiz. **Desired:** * Moderately concentrated European spatial organiz.	**Expected:** * Large impact fuel price increase * Mixed European spatial organiz.
41-50	**Expected:** * Low profitability transport * Much car use		
> 50	No large differences		

The resulting expected scenario is presented in Figure 11.8. The scores in the transport system, however, do not differ greatly from the average scores, so that an entirely different transport system is not expected from the various developments in the distinct fields.

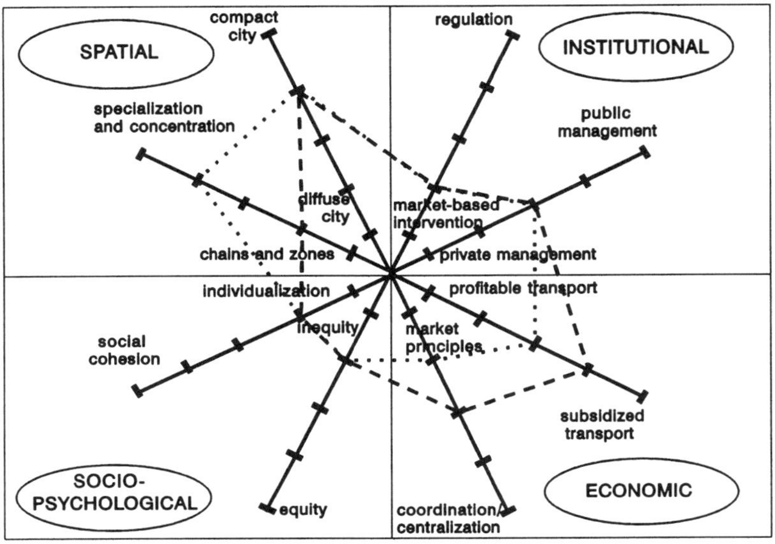

Figure 11.8. Expected scenario of ministry respondents, age cohort 41-50 years

Notes: 1) $n = 16$

2) .... average; -------- ministry, 41-50 years

Municipality, 41-50 years

This subcategory puts more emphasis on using collective modes, especially in the expected scenario. This trend is reflected in the high scores for behavioural options, policy measures and distinct modes in both the expected and desired transport system. Also less emphasis on the profitability of the railways is expected and desired. On the other hand, the expected spatial European organization is moderately diffuse, which is not entirely consistent with an emphasis on collective modes; a mixed development appears to be desired, however. This gives the scenarios as presented in Figures 11.9 and 11.10.

The resulting transport system places - especially in the expected scenario - more emphasis on collective modes, while also road pricing has a high score.

From the new modes especially the electric car gets a high score; the other modes do not have higher scores in most cases.

Consultancy, < 30 years
It is interesting that for the group of young consultants the expected scores for most policy measures which reduce car use or stimulate collective modes use are relatively high, while the desired scores are lower than the average. Therefore, their expected policy is like their desired policy. In both the expected and the desired scenario the profitability of the railways is valued low; the same holds for road pricing. Therefore, the total profitability of the transport sector is rather low.

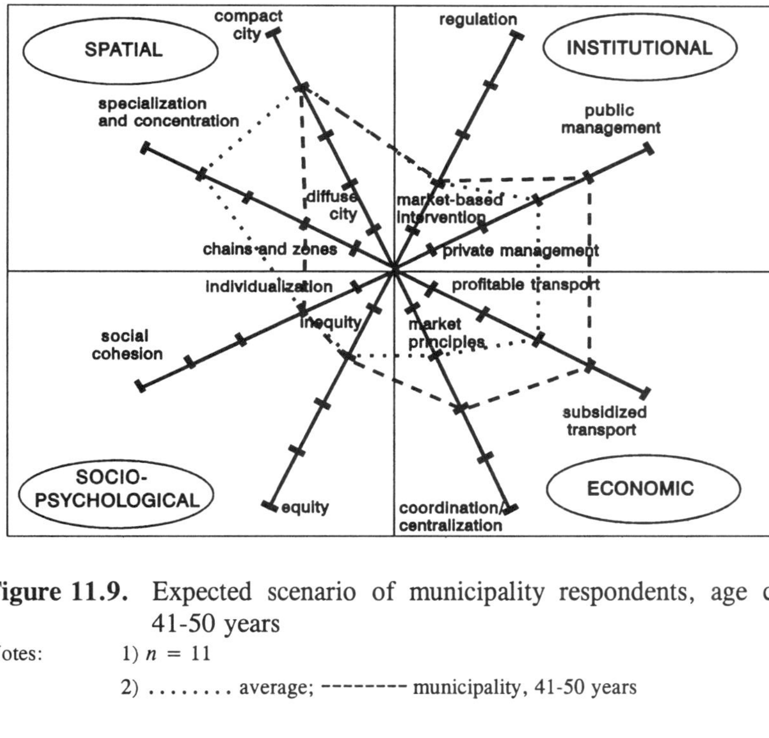

Figure 11.9. Expected scenario of municipality respondents, age cohort 41-50 years

Notes: 1) $n = 11$
2) .... average; -------- municipality, 41-50 years

Slightly different answers from the average are given for the desired level of decision making, which should be at the national and regional level. The individualization trend is expected to stop, while the desired urban spatial organization is moderately compact. The resulting scenarios are presented in Figures 11.11 and 11.12.

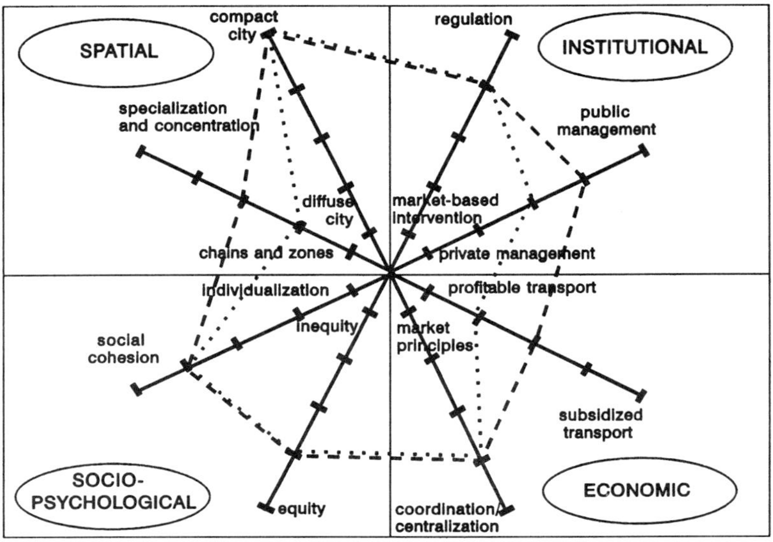

Figure 11.10. Desired scenario of municipality respondents, age cohort 41-50 years

Notes: 1) $n = 11$

2) .... average; -------- municipality, 41-50 years

The resulting transport system - according to the average scores for the distinct modes - means that the expected system has low scores for many collective modes, especially in the urban system. Road pricing in particular is valued very low for all submarkets. The scores for the desired system are about the same as the average, although the Maglev high-speed train and subterranean shuttles especially get relatively high scores.

Consultancy, 41-50 years

This subcategory puts less emphasis on the use of collective modes. In addition, rising fuel prices are favoured and regulatory measures get relatively low scores in the expected scenario. It is also expected that profitability increases of the railways will be low, while the expected spatial organization is moderately diffuse and the level at which transport decisions are taken is at the regional level. The resulting expected scenario is presented in Figure 11.13.

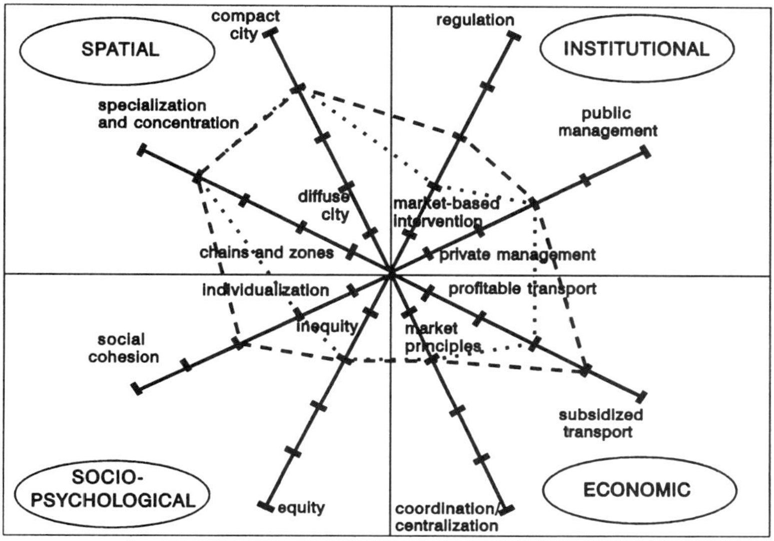

Figure 11.11. Expected scenario of consultancy respondents, age cohort < 30 years

Notes: 1) $n = 13$

2) .... average; -------- consultancy, < 30 years

In all submarkets the expected transport system places less emphasis on collective modes, although road pricing will be introduced to a somewhat greater extent. The private car has a relatively large market share. In the desired scenario the scores are more or less the same as the average.

University, 41-50 years

The views of this group of Dutch transport experts associated with a university also lead to some interesting findings. Most of the desired - and to a lesser extent the expected - behavioural options all get lower scores. So in these scenarios there is less reduction of car use and less increase of collective mode use. Also all policy measures get low scores. It is interesting also that little impact on car use is expected from increasing fuel taxes. In the desired scenario also the profitability of the railways is lower than the average, while main decisions are taken at the regional level. The resulting scenarios are presented in Figures 11.14 and 11.15.

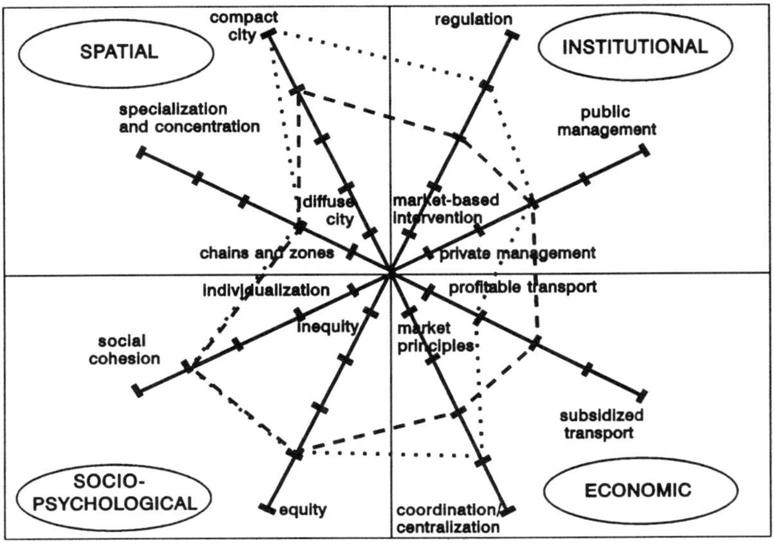

Figure 11.12. Desired scenario of consultancy respondents, age cohort < 30 years

Notes: 1) $n = 13$

2) .... average; -------- consultancy, < 30 years

In the resulting transport system road pricing gets low scores in the urban and regional system. The same holds for collective modes, while (improved) cars get high scores. So the resulting transport system is in both the expected and desired scenario more individual than in the average expert scenarios.
The desired profitability of the railways is rather high. The desired transport system is about equal to the average scenarios, but only in the desired long-distance system is more emphasis put on collective modes.

The main differences between the general scenarios and the subgroups with respect to age and profession are depicted in Table 11.3. They are related with the use of public transport. Several subgroups expect the profitability of railways to be low, while also differences are found with respect to the use of collective transport modes: some subgroups expect a higher, others a lower use of these modes. Also in the spatial field some interesting differences are found: several subgroups indicate a more diffuse spatial organization at both the European and urban level. Interestingly, there is almost no discrepancy in expectations and desires in the economic, institutional and social-psychological field. Again, the agreement among the experts appears to be high.

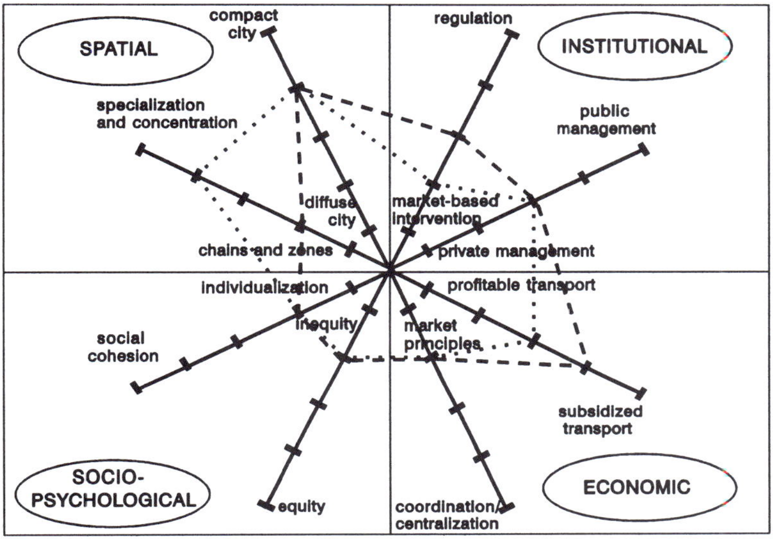

Figure 11.13. Expected scenario of consultancy respondents, age cohort 41-50 years

Notes: 1) $n = 14$

2) .... average; -------- consultancy, 41-50 years

In conclusion, the age cohort analysis explains some variations in the findings of different subgroups. Although such differences are relevant, they do not lead to the need or the possibility to generate entirely new scenarios.

11.4 MINORITY SCENARIOS

The expert scenarios described in Chapter 10 have been constructed by using the highest frequencies and the average scores per question and option. It is still possible, however, that there are substantial minorities giving alternative answers, while also the average scores may hide some 'peaks' or 'valleys' in the scores given.

When the frequencies of the answers are analysed, we observe in particular an inconclusive result for the European spatial organization. The main minority answer in the expected case is moderately diffuse, while in the desired organization this is moderately compact. For the modal split a large minority expects a reduced modal share of cars. A large minority expects that there will be no clear administrative level where transport decisions are taken, while such decisions should desirably be taken at the regional level.

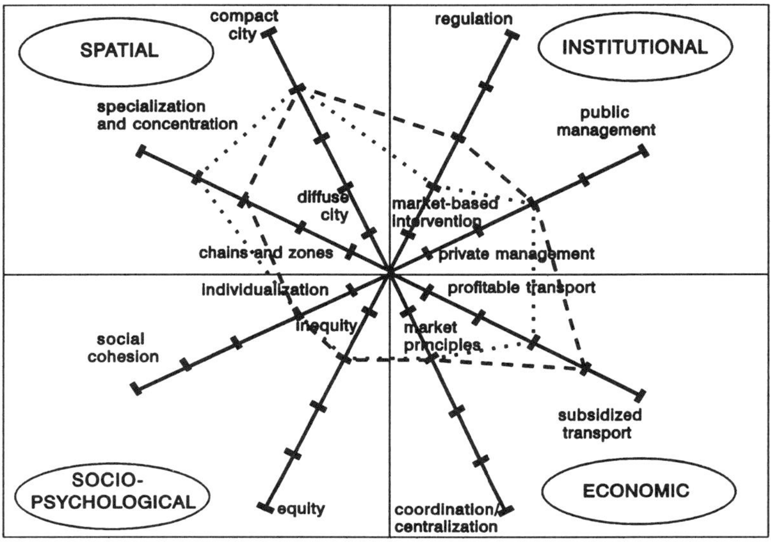

Figure 11.14. Expected scenario of university staff, age cohort 41-50 years

Notes: 1) n =11

2) .... average; -------- university, 41-50 years

The individualization trend may come to a halt according to a large minority. The scores for behavioural and policy options are rather equally distributed around the average. The same holds for the profitability scores in both the expected and desired scenario. By putting together all such minority views in one Spider, the resulting scenarios, where at relevant axes large minority answers are observed, can be presented as shown in Figures 11.16 and 11.17.

The scores for the different modes in the transport system are in general equally distributed around the average scores. Exceptions are road pricing in urban areas, which is valued only with a '1' by a large minority (on a scale from 1 to 10, 1 is the lowest and 10 is the highest), while also many value this option with a '7' or even higher in both the expected and desired system.

The same high score holds for cycling in regional transport (many apparently expect large improvements in bicycles which make them attractive for longer distances), car pooling in the expected long-distance system, and the Maglev in the desired system. It is also interesting that the desired scores for the conventional car on long distances are valued lower than 6 by about 50 per cent of the respondents, while also about 50 per cent value this option higher than 6.

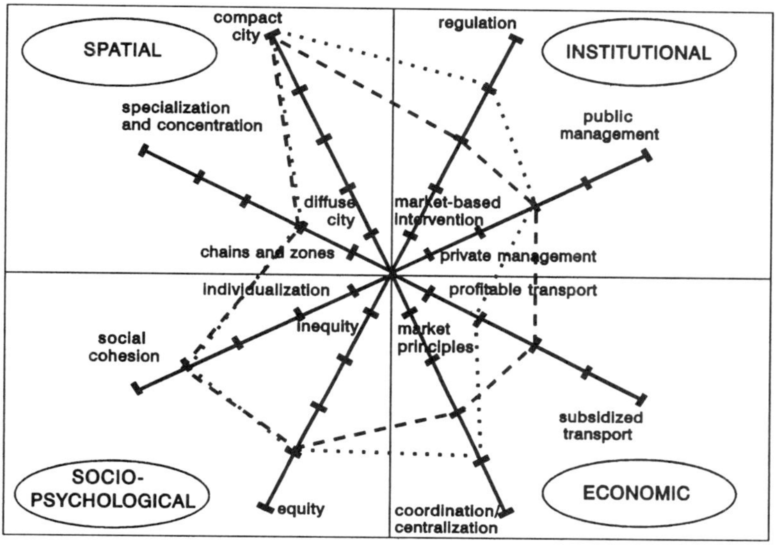

Figure 11.15. Desired scenario of university staff, age cohort 41-50 years

Notes: 1) $n = 11$

2) .... average; -------- university, 41-50 years

In general, however, no clear picture of an alternative transport system is found when we take the scores of large minorities into account, although the scenario leading to it is a slightly different one from the average.

11.5 CONCLUSIONS

It may be concluded with regard to the analysis per subcategory of experts and of large minorities that the scenarios do not differ to a large extent. Therefore, we come to the interesting conclusion that there is apparently much agreement among the Dutch transport experts about the future of transport. An explanation for this may be the small size of the country, the easy access to all transport information, the social cohesion or communication among these experts and the frequent exchange of any relevant information in research papers, articles and also at regular meetings of transport experts.

Moreover, it is also noteworthy that most of the views of the Dutch experts are shared by their European colleagues, as was shown in Chapter 10. However, the interesting question remains whether the statements on the degree of

Table 11.3 (a). Main scenario differences of subsegments of respondents by profession and age

	< 30	30 - 40
Ministry	**Expected:** * Moderately diffuse European spatial organization **Desired:** * Individualization reversed	**Expected:** * Low profitability railways * Centralization
Municipality		**Desired:** * Decentralization
Consultancy	**Expected:** * Less car use * Low profitability railways * Individualization stops **Desired:** * Decentralization * Low profitability railways * Mod. compact urban spatial organization	**Expected and desired:** * More new fuel and electric cars
University		
Public transport company		**Expected:** * Mixed urban spatial organization * More use collective modes **Desired:** * More regulation * More use collective modes

sustainability of each of these scenarios can be further underpinned by empirical and testable information, in particular regarding CO_2 emissions. This issue will be dealt with in the next chapter.

Table 11.3 (b). Main scenario differences of subsegments of respondents by profession and age

	41 - 50	> 50
Ministry		
Municipality	**Expected:** * More use collective modes * Low profitability railways * Moder. diffuse European spatial organization **Desired:** * Low profitability railways * Mixed European spatial organization	
Consultancy	**Expected:** * Less use collective modes * Less regulation * Low profitability railways * Decentralization	
University	**Expected:** * More car use * Less regulation **Desired:** * Less car use * Low profitability railways	
Public transport company	**Desired:** * Centralization	**Expected:** * Centralization **Desired:** * Low profitability railways * More use collective modes

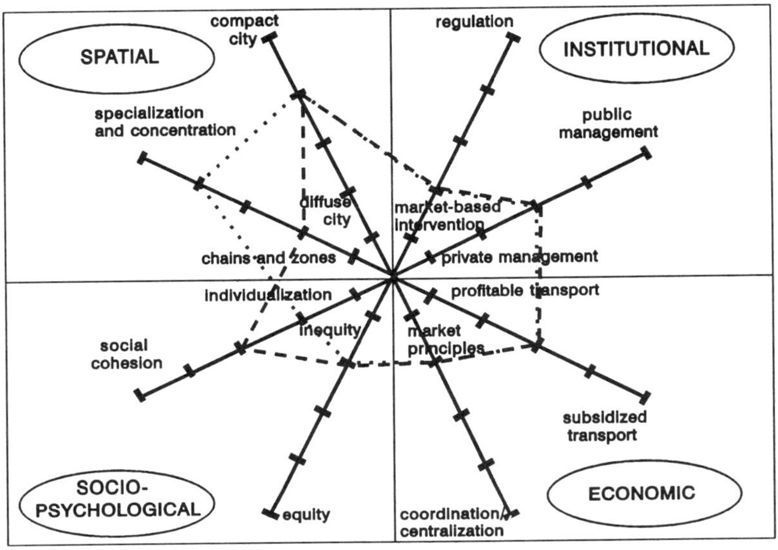

Figure 11.16. Expected scenario with the opinion of large minorities on relevant axes

Notes: 1) n =271 2) .... average; -------- large minorities

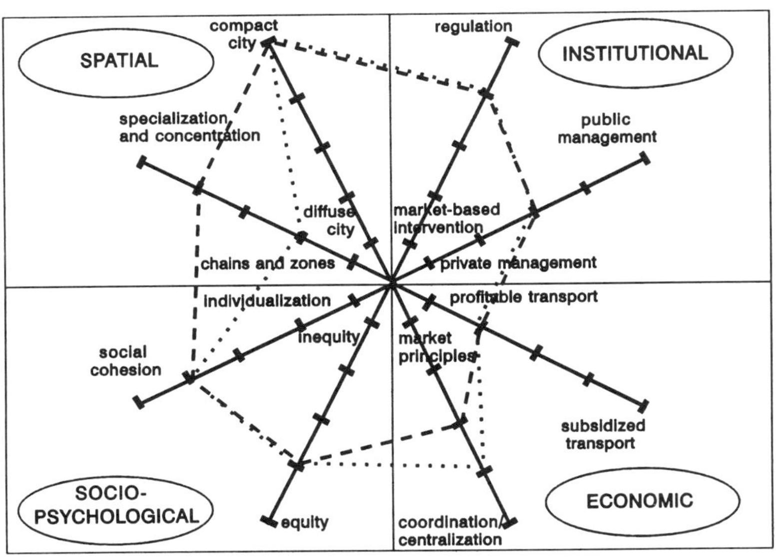

Figure 11.17. Desired scenario, idem. Notes: See Figure 11.16

12 ENVIRONMENTAL IMPACTS: ESTIMATION OF CO_2 EMISSIONS

12.1 INTRODUCTION

Chapters 8 and 10 four different scenarios (two reference scenarios and two expert scenarios) have been qualitatively described. These scenarios sketch different development paths of the spatial, institutional, socio-psychological and economic determinants of passenger mobility, as well as offering a general description of the resulting transport system. From these descriptions it cannot be inferred in a straightforward way whether these developments lead to reductions in, for example, the volume of CO_2 emissions. This requires a more thorough analysis of the spatial pattern, the modal split, the behavioural responses and the transport technologies involved. Therefore, this chapter provides quantitative estimations of the emissions in each scenario. In this way, it can be tested whether the environmental implications of each of these transport scenarios fulfils the targets on CO_2 reduction as set by governments.

For each scenario we will sketch how the passenger transport system might look in terms of the use of various existing and future transport technologies and the corresponding modal split. Expected energy consumption characteristics of the various transport modes, figures from the present fuel supply and electricity generation system, as well as estimations of the future energy system will be used. Next, we estimate the energy consumption and CO_2 emissions by the future passenger transport systems and compare these impacts with the current systems. The present transport system is described and analysed on the basis of available European statistics for 1989 (ECMT, 1993; Eurostat, several years). This means that some average European assessments can be made.

Seventeen European countries are included in the calculations: Austria, Belgium, Denmark, Finland, France, Germany, Greece, Ireland, Italy, Luxembourg, the Netherlands, Norway, Portugal, Spain, Sweden, Switzerland and the United Kingdom. For each of these countries the transport system and the energy supply system, the energy consumption and CO_2 emissions are assessed. In the final results the figures of individual countries are added up to obtain European averages.

It should be emphasized that the various estimates are tentative. Input data describing future overall passenger mobility and the future composition of the energy system in the four different transport development scenarios are dependent on the assumptions made, although available studies and forecasts are used as far as possible. An advantage of this approach is that several energy consumption pathways can be used and tested; the same may hold for different mobility levels. As a consequence, the results are mainly intended to give an indicative picture of the possible impacts on sustainability as a result of the developments described in the reference and expert scenarios. These results also aim at describing the impact of several energy policies.

Section 12.2 deals with the energy characteristics of present and future transport technologies. Section 12.3 describes the present and future electricity generation and fuel supply systems. The mobility rate and modal split are described in Section 12.4. In Section 12.5 the data found are combined to calculate the overall energy consumption and CO_2 emissions per scenario. Finally, in Section 12.6 some conclusions are drawn.

12.2 ENERGY CONSUMPTION AND CO_2 EMISSIONS OF TRANSPORT MODES

As mentioned above, the sustainability of transport systems will be tested by confronting their CO_2 emissions with generally accepted target levels of CO_2 reduction. The first step in calculating the CO_2 emissions of transport modes is to estimate their energy consumption. In the scenarios, a distinction is made between urban and long-distance passenger transport. The energy consumption of transport modes is expressed in terms of the energy content of final energy carriers (electricity or motor fuels like gasoline, diesel, biofuel, kerosene or hydrogen) consumed per passenger kilometre.

Due to the aggregated nature of the available European transport data, only four transport modes are discerned in the 1989 transport system: the car, bus, train and plane. Table 12.1 provides the final energy consumption factors of these modes for urban and long-distance passenger transport in 1989.

Table 12.1. Final energy consumption of various transport modes (1989, MJ/pass.km.)

Car		Bus		Train		Plane
Urban	l.d.[1]	Urban	l.d.	urban	l.d.	l.d.
1.70	1.17	0.75	0.60	0.35	0.35	2.00

Note: 1) l.d. = long distance
Sources: Smokers and Van Hilten (1992); IEA/OECD (1993); confidential ECN-data

As the estimates of a specific technology often vary from source to source, the data specified in this section are, in general, averages, weighted on the basis of expert judgement and expert estimations. In most cases the specific numbers cannot be related to a single literature source, and are therefore presented without reference.

For the future scenarios in our study a larger set of transport technologies has to be taken into account. Table 12.2 presents the estimated final energy consumption of these transport modes in the year 2030, in which five different passenger cars and buses are considered in the calculations:

- a very energy efficient version of the conventional car (internal combustion engine vehicle (ICEV));
- an ICEV with a biomass-derived alternative fuel (e.g. methanol);
- a hybrid-electric car (HEV), with an electric drive system and a gasoline-fuelled combustion engine operating as generator;

Table 12.2. Expected final energy consumption of transport modes (2030, MJ/pass.km)

Urban transport		Long-distance transport	
Car		**Car**	
ICEV conv.	0.90	ICEV conv.	0.70
ICEV alt. fuel	0.90	ICEV alt. fuel	0.70
HEV	0.50	HEV	0.55
BEV	0.40	Fuel cell	0.45
Bus		**Bus**	
ICEV conv.	0.50	ICEV conv.	0.55
ICEV alt. fuel	0.50	ICEV alt. fuel	0.45
HEV	0.40	HEV	0.35
Fuel cell	0.40	Fuel cell	0.30
BEV	0.20		
Tram	0.25	**Train**	0.30
Metro	0.25	**HST**	0.35
People Mover	0.30	**Plane**	
		Conventional	1.40
		LH_2	1.20

- a fuel-cell powered car (with hydrogen as energy carrier);
- a battery-electric car (BEV) (only for urban transport).

In urban public transport trams, metros and people movers are used, which are all electrically driven. Collective long-distance transport, too, is performed by electrically propelled conventional trains, high-speed trains (HST) and trains with (electro-)magnetic levitation and propulsion (Maglev). Planes may be conventionally (kerosene) fuelled planes, or liquid hydrogen (LH_2) fuelled (see also Chapters 5 and 6).

Although the technological development of future transport modes will be influenced by the spatial, institutional, economic and socio-psychological aspects that also determine the future development of mobility, we use the same energy consumption figures for every mode in each scenario. This is because the four scenarios significantly differ in their modal split. Generally, it is not useful, therefore, to describe the technological state of development of a certain transport technology, when it is hardly used. As a result, it is assumed that the energy characteristics of transport modes will develop favourably under the conditions of the scenarios in which they are extensively used. In all scenarios, therefore, relatively optimistic energy characteristics can be used.

12.3 ELECTRICITY AND MOTOR FUELS

After the description of the various modes in terms of energy consumption it is necessary to investigate which fuels are used and the emissions of these energy sources. In this respect we have to take the whole energy chain from primary energy sources (coal, gas, oil, uranium, etc.) to final energy sources (electricity and fuels) into account.

First, we will discuss the electricity generation as well as the assumptions made in the calculations for the scenarios. Next, the figures on future motor fuels will be dealt with.

12.3.1 Electricity generation

The scenarios contain different assumptions for the generation of electricity. These assumptions have large impacts on the final CO_2 emissions. In order to calculate the primary energy consumption of electrically driven modes, their final energy consumption factor must be divided by the efficiency with which it is generated. In a similar way, the resulting CO_2 emission per passenger kilometre is calculated by multiplying the final energy consumption with a CO_2 emission factor. The efficiency of the electricity generation system is calculated by dividing the yearly electricity output by the yearly input of primary energy carriers.

Similarly, the CO_2 emission factor is defined as the total yearly CO_2 emission from the electricity generation divided by the total yearly electricity production. The conversion efficiencies and CO_2 emission factors of electricity generation from coal, gas, oil, uranium and renewables are presented in Table 12.3.

Table 12.3. Characteristics of processes for electricity generation

Energy source	Coal	Gas	Oil	Nuclear	Renewable
g CO_2 per MJ prim.	94	56	74	0	0
1989					
Efficiency (%)	36	37	36	33	100
g CO_2 per MJ fin.	260.39	152.59	208.45	0.00	0.00
2030					
Efficiency (%)	50	58	40	33	100
g CO_2 per MJ fin.	188.00	96.55	185.00	0.00	0.00

For each of the above-mentioned 17 European countries (see Section 12.1) the composition of the electricity generation system is projected for the scenarios. The input for the expected scenario is based on the average European energy data for the year 2020, as estimated by Oostvoorn and Pellekaan (1995). The composition of the electricity generation in the other three scenarios is derived from that of the expected scenario in an intuitive way, taking into account the distinctive characteristics of the institutional, economic and socio-psychological determinants in the scenarios. It has been attempted to create four distinctively different energy scenarios in order to broaden the analysis and to analyse the sensitivity of the outcomes. The resulting shares of various energy sources in the European electricity generation mix are presented in Table 12.4.

Next, the assumptions made in the distinct scenarios will be elaborated.

The electricity generation in the market scenario

The market scenario is characterized by private investment and resulting profitability goals, combined with market-oriented government intervention by means of a CO_2 tax. Despite this tax private investment in nuclear energy does not seem very likely. The use of nuclear energy is therefore less than in the expected scenario. The use of renewables (hydroelectric power, photovoltaics (PV) and wind energy) on the other hand is much higher than in the expected scenario; because of the CO_2 tax these sources are profitable. The use of small- and medium-sized gas-fired power plants (cogeneration) is expected to be favoured by private investors. Despite the CO_2 tax the use of electricity from coal is still significant, because of the low costs of coal.

Table 12.4. Share of various energy sources (per cent energy input)

Scenario \ Source	Coal	Gas	Oil	Nuclear	Renewable
1989	35	6	9	38	12
Market	22	16	7	27	28
Regulatory	8	10	3	48	31
Expected	30	7	13	38	12
Desired	24	13	5	21	37

The electricity generation in the regulatory scenario

The regulatory scenario is characterized by strong government intervention by means of, for example, strong CO_2 emission reduction targets, while the government is responsible for all large-scale infrastructure. In this scenario it is assumed that the European governments invest strongly in nuclear energy to realize CO_2 emission reductions in the short term. Next-generation nuclear technology is used, which has more inherent safety than the present nuclear generation reactors. European governments are also assumed to invest in large-scale renewable energy projects (especially hydroelectric power and photovoltaics). This results in a strong decline in the use of coal in comparison with 1989 and the expected scenario.

The electricity generation in the expected scenario

The expected scenario can be seen, to some extent, as a 'business as usual' scenario. For this scenario energy data from Oostvoorn and Pellekaan (1995) have been used. Although there are significant shifts in the energy mix of almost all European countries, the average composition is more or less equal to the composition in 1989.

The electricity generation in the desired scenario

The desired scenario is - like the regulatory one - characterized by strong government influence. However, in contrast to the regulatory scenario large-scale investments in nuclear energy are considered to be socially unacceptable. The use of nuclear power has, therefore, declined by 50 per cent compared to the expected scenario. This is partly compensated by the extensive use of renewable energy sources but the drastic reduction of nuclear generation makes it inevitable to use coal and gas for a significant part of the electricity production.

12.3.2 Motor fuels

Table 12.5 shows some characteristics of the motor fuels that are taken into account in the estimations for the base year 1989.

Table 12.5. Characteristics of various motor fuels (1989)

	MJ fin/ MJ prim. (%)	MJ fin/ MJ fossil (%)	CO_2 prim. (g/MJ)	CO_2 fuel (g/MJ)	CO_2 fin. (g/MJ)
gasoline	90.00	90.00	74.50	74.50	82.78
diesel	93.00	93.00	73.00	73.00	78.49
kerosene	90.00	90.00	73.00	73.00	81.11

Table 12.6. Characteristics of various motor fuels (2030)

	MJ fin./ MJ prim (%)	MJ fin./ MJ fossil (%)	CO_2 prim. (g/MJ)	CO_2 fuel (g/MJ)	CO_2 fin. (g/MJ)
Gasoline	90.00	90.00	74.50	74.50	82.78
Diesel	93.00	93.00	73.00	73.00	78.49
Alt. fuel[1]	55.00	1.0 E+15	0.00	70.00	0.00
Fuel cell[2]	100.00	1.0 E+15	0.00	0.00	0.00
Kerosene	90.00	90.00	70.00	70.00	77.78
LH_2[2]	100.00	1.0 E+15	0.00	0.00	0.00

Notes: 1) bio-ethanol
2) H_2 from PV

The first two columns present the efficiency of fuel refining as units of final energy input divided by units of primary or fossil energy input, which are equal for all fossil fuels. The third and fourth column give the CO_2 emission per unit of primary and final energy consumed, the last column the CO_2 emission per unit of fuel energy content. In the CO_2 emission per unit of final energy consumed the indirect emissions resulting from oil refining have been incorporated too.

Similar characteristics of the motor fuels related to the year 2030 are presented in Table 12.6. Besides gasoline, diesel and kerosene, two alternative motor fuels have been considered in our analysis. These are a biomass-derived fuel (e.g. bio-ethanol or biodiesel) and hydrogen. In road transport hydrogen is assumed to be used in fuel cells. Hydrogen can also be burnt in the gas turbines of aircraft, which carry the fuel in liquid, cryogenic form. The efficiency of the production of alternative fuel from primary biomass is relatively low. It is assumed that all energy requirements in the energy chain of biomass derived fuel are covered by energy from biomass. The efficiency of fossil to final energy conversion is, of course, infinite for biomass-derived alternative fuel and hydrogen, but has been set to a sufficiently high finite number in the model calculations.

12.4 DEVELOPMENTS IN PASSENGER MOBILITY

For the estimation of the CO_2 emissions the mobility level in the distinct scenarios also has to be determined. This will be dealt with in the next subsections. First, the mobility level in 1989 is investigated; next, the assumed mobility levels in the distinct scenarios are discussed.

12.4.1 The mobility level in 1989

The passenger mobility and modal split in European countries in 1989 is not easy to measure, but it can be estimated by using ECMT (1993) and Eurostat (several years) data. For Greece, Ireland and Luxembourg complementary data have been found in Samaris and Zierock (1992). Available statistics do not make a distinction between passengers in urban and long-distance transport, however. Since such a distinction is necessary in the scenarios, the division of total mobility into urban and long-distance transport has been estimated by taking the trips below 10 kilometres as urban and above 10 km as long-distance transport and estimating their average length. Although this assumption may be rather crude, it is necessary to include all passenger kilometres in the analysis. Moreover, the results as presented in Table 12.7 look plausible, and in any case this assumption is not expected to have a large impact on the final results.

Because it was impossible to calculate the distance distinction for all European countries separately, a European average is used for all countries. The total European passenger mobility in 1989 amounted to 3948 x 10^9 pass.km of which

27 per cent is assumed to be urban transport and 73 per cent long-distance transport.

Starting at this point, the mobility level in the scenarios can now be estimated. This will be worked out first by extrapolating current trends and by discussing the assumptions made in the successive scenarios.

Table 12.7. Division of passenger transport in Europe (1989)

	Car		Bus		Train		Plane
	urban	l.d.	urban	l.d.	urban	l.d.	l.d.
European average	29	71	50	50	10	90	100

Note: l.d. = long distance

12.4.2 Trends in mobility

In most studies about future transport scenarios a simple model is used, in which current trends in specific parameters are translated into the future by means of a linear multiplier. This assumes that past trends may be used as an indicator for future trends. If we apply this to developments in Europe, the expectations may be as presented in Table 12.8.

If the past trends are to be regarded as a blueprint for the future, mobility will apparently explode. However, one may doubt whether this will actually happen because in these trends it is assumed (see also Section 3.3.2) that:

- there is no restriction in the capacity of the infrastructure; in reality, however, unlimited infrastructure construction is not realistic;
- the time budget of individuals is unlimited; in practice, however, this is not realistic (of course faster modes may be introduced);
- car ownership should grow at the same rate as the last 25 years; however, this neglects the existence of saturation;
- population growth should continue at the same pace.

There are no applicable population forecasts available for 2030. The major reason for this is the way in which population data are gathered in the various European countries. For the EU, Eurostat (1993) includes, for example, a forecast for only seven countries for the year 2020. This forecast indicates the population will stabilize at the beginning of the next century, which is mainly due to the low birth rates.

For the above-mentioned reasons it may be expected that current mobility growth will continue at a slower rate. In order to find more plausible growth rates, therefore, we will discuss some Dutch forecasts, which may be an indication of the European growth rates.

Table 12.8. European mobility trends

	Av. Growth rate 1965-1989 (%)	Index 2030 (1989 = 100)
Total	3.7	428
Car	4.3	539
Train	1.1	155
Bus	1.3	168
Air	8.1	2254

Notes: See Figure 3.2
Sources: Own calculations based on ECMT (1993) and Eurostat (several years)

12.4.3 Dutch mobility predictions

According to the official Dutch transport policy memorandum (Tweede Kamer, 1990) car mobility for the year 2010 will rise to some 70 per cent above the 1986 level (without additional restraining policy measures). This means an average growth of 2.15 per cent per year, which is considerably lower than the actual growth in the period 1965-1989 (4 per cent per year), which clearly followed the European trend. For these mobility growth predictions no separate distinction can be made per mode.

A more recent study by the Dutch CPB (1992b; see also Section 3.7.1), describes three major socio-economic scenarios for the year 2015 and their consequences for transport. In Pronk *et al.* (1992) these scenarios have been used for an analysis which includes various policy packages for the same year. According to their calculations, mobility rise in the year 2015 will be 60 to 91 per cent higher than 1990 level. This means average growth rates of between 1.9 and 2.6 per cent per year, which is about the same as found in Tweede Kamer (1990).

Based on the foregoing observations and assuming no drastic policy changes - only some policy measures to reduce mobility growth - we may expect some reduction in the growth of mobility for the period up to the year 2030.

12.4.4 The mobility rates in the scenarios

By taking the above considerations into account, we have estimated the mobility level in the distinct scenarios. Table 12.9 presents the growth indices of urban and long-distance passenger transport in the different scenarios distinguished (1989 = 100). The assumptions made in these scenarios will be discussed next.

Table 12.9. Growth indices in passenger transport (1989 = 100)

Scenario	Market		Regulatory		Expected		Desired	
	urb.[1]	l.d.[2]	urb.	l.d.	urb.	l.d.	urb.	l.d.
European average	125	200	185	130	185	225	150	185

Notes: 1) urb. = urban
2) l.d. = long distance

The mobility level in the market-oriented scenario

In the market-oriented reference scenario mobility growth is mainly concentrated in the high income group, which consists of 20 per cent of the population, owns a private car and mainly lives in peripheral regions. Also some mobility growth in the middle income group is likely to occur.

When we analyse the assumptions made in Chapter 8, the following picture emerges. In long-distance transport, mobility growth in the 2000s is initially negative. If we assume that in the 1990s the growth in mobility was 3 per cent per year, then the 2000 mobility level may lie 40 per cent above the 1989 level. After the year 2000, mobility growth is reduced to 30 per cent above this level. After 2010 mobility grows mainly without restrictions, though it is more expensive than in the 1990s. Therefore, the growth rates are expected to be lower and to rise in the year 2030 to be 100 per cent above the 1989 level, which means an average growth rate of 1.7 per cent per year.

The urban picture differs considerably from the long-distance one. Although the 2000 level equals that of long-distance traffic, this level will stabilize afterwards because of the decreasing population in urban areas. Therefore, this level is expected to be 25 per cent higher than the 1989 level (or a growth rate of 0.5 per cent per year).

The mobility level in the regulatory scenario

In the second reference scenario, all major government policies aim at redressing mobility growth by regulatory measures. The use of individual modes is especially reduced, whereas the use of collective modes has become more attractive due to an increase in investment outlays and government support for collective transport operators. As a result, the decline in individual mobility is offset by the increase in collective transport.

In long-distance transport, the overall mobility level may be relatively low because of the shift towards urban areas and the lower quality (more time-consuming) transport modes. In the 2000s, the mobility level even declines because of the policy measures implemented by then. Therefore, this level may in 2030 be some 30 per cent above the 1989 level, which implies a growth rate of 0.6 per cent per year.

Because of the shift back to the urban areas and the relatively high quality of the collective modes, the mobility level will grow much more in the urban areas, viz. 85 per cent (on average 1.5 per cent per year).

The mobility level in the expected scenario

In the expected expert scenario, it is taken for granted that in the period up to the year 2030 mobility growth will continue, although at a slower pace than in the period before the year 1989, for the above-mentioned reasons. It may be assumed that long-distance mobility will grow at a rate of 2 per cent per year, which means that the 2030 mobility level lies 125 per cent above the 1989 level.

Also, in the urban areas we will observe a considerable mobility increase, which is, however, lower than in long-distance transport, because of the development of the moderate compact city, which reduces transport needs. Therefore, the growth rate is 1.5 per cent per year, which means that the mobility level rises to 85 per cent above the 1989 level. In total, the resulting mobility growth is lower than in the official Dutch future predictions, because it is assumed that the reduction in growth rates will continue.

The mobility level in the desired scenario

Interesting in the desired expert scenario is the fact that many long-distance criss-cross relationships are covered by collective modes, which decreases the efficiency of transport. Therefore, the mobility growth rates are lower than in the expected case: 1.2 per cent per year, which is, however, higher than in the regulatory scenario.

At the urban level the compact city concept is introduced in this scenario.

After the introduction of anti-car measures, mobility declines quite strongly, but later the level will rise again due to economic and population growth and the occurrence of new services. Therefore, the former growth rate is re-established. In total, this results in a mobility level of 50 per cent above the 1989 level or an average growth rate of 1 per cent per year. This is lower than in the regulatory scenario, because there the urban population growth is lower than in the desired scenario. In the the broader European mobility analysis, identical growth rates are assumed for all countries. By applying these growth rates to the total urban and long-distance mobility of individual countries in 1989, the urban and long-distance mobility in the scenarios can be calculated. The total passenger mobility in 1989 and for the scenarios for 2030 is depicted in Figure 12.1.

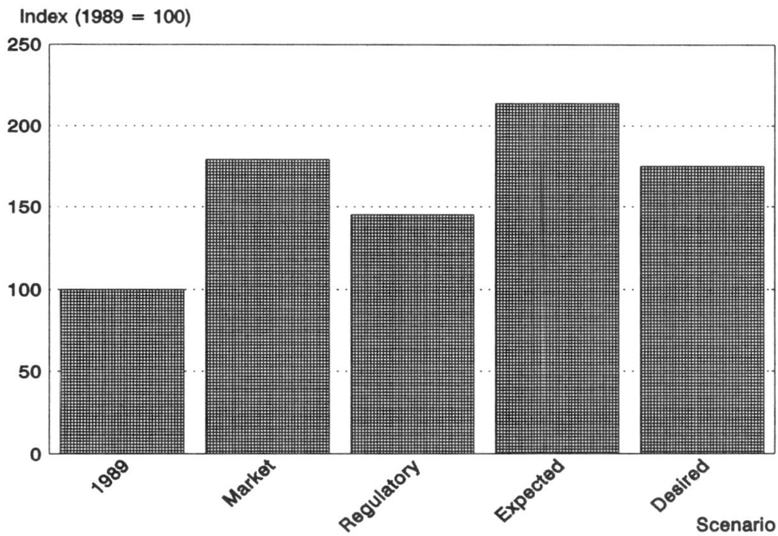

Figure 12.1. Estimated mobility level in the scenarios (index: 1989 = 100)

The total mobility in the expected scenario has apparently more than doubled compared to 1989 (115 per cent increase). In the market-oriented scenario overall mobility has increased by 82 per cent compared to 1989. The overall mobility is lowest in the regulatory scenario, with an increase of 46 per cent compared to 1989. Finally, the desired scenario has more or less average mobility levels, with an overall mobility of 77 per cent above the level of 1989.

12.4.5 The future modal split including new technologies

Given the overall mobility data for urban and long-distance passenger transport in the scenarios, we will now describe the consequences of implementing the various transport technologies. The scores as presented in Chapter 9 are used as main indicators, while the other characteristics of the scenarios - in terms of the spatial, institutional, economic and socio-psychological determinants - have been used as a basis for composing a conceivable modal split for each scenario. When a mode is supposed to almost disappear or is not introduced in a scenario, the modal share is assumed to be '0'. In this way the scenarios have been made more distinct from one another in order to focus the scenario analysis. The assumed modal splits in terms of the shares of various technologies in urban and long-distance transport are presented in Table 12.10 and Table 12.11.

Both reference scenarios present a clear modal split in favour of either the car or collective modes, while the expert scenarios are somewhat less extreme. The modal share of the plane stays low in each scenario, a situation also indicated in the scores for the distinct modes. For the modal split between car types, the assumptions made in Section 12.3.1 for electricity generation are taken into account. All these data are now used to assess the energy consumption for the different scenarios.

12.5 ANALYSIS RESULTS

12.5.1 General results

The total number of passenger kilometres in each of the European countries has been calculated for all transport technologies in 1989 and 2030. By using the above-discussed data, the total consumption of final, primary and fossil energy and the CO_2 emissions by European passenger transport in 1989 and in the scenarios for 2030 can now be calculated (see Figure 12.2 and 12.3). The big difference in fossil energy use between the expected scenario and the other scenarios is striking. The same holds for CO_2 emissions.

Figure 12.4 presents the relative consumption of various primary energy carriers by European passenger transport in 1989 and in the scenarios for 2030.

Actually, the hydrogen (H_2) listed in Figure 12.4 is not a primary energy carrier. In all scenarios hydrogen is assumed to be produced from renewable energy sources (e.g. photovoltaic) with 100 per cent (virtual) efficiency, however. The contribution of hydrogen fuel has been split off here from the total contribution of renewables in order to be able to distinguish it from the use of renewables through electricity consumed by electrically propelled transport systems. The results for the individual scenarios will be described and analysed in the next subsection.

Table 12.10. Urban modal split in the scenarios (%)

Mode \ Scenario	Market	Regulatory	Expected	Desired
Car				
ICEV conv.	0	0	20	0
ICEV alt. fuel	35	0	10	10
HEV	0	0	0	0
BEV	40	10	10	10
Total car	75	10	40	20
Bus				
ICEV conv.	0	0	15	0
ICEV alt. fuel	10	0	0	0
HEV	0	10	10	10
Fuel cell	5	15	5	10
BEV	10	10	10	10
Tram	0	20	10	20
Metro	0	20	5	20
People mover	0	10	0	10
Train	0	5	4	0
Total coll. modes	25	90	60	80
Total	100	100	100	100

12.5.2 Analysis of results for the individual scenarios

The market scenario

Figure 12.4 shows that in 1989 the passenger transport system almost completely relied on the use of oil as a primary energy carrier. This is, of course, the result of the dominant position of the passenger car. In the market scenario this dominant position of individual road transport has not changed. From Tables 12.10 and 12.11 it can be seen that the car accounts for 75 per cent of all urban passenger transport and 90 per cent of all long-distance transport. The use of oil, however, has almost completely been substituted with biomass-derived motor fuels. In total, yearly consumption of primary biomass-energy amounts to 7.5 EJ. According to Okken *et al.* (1993a) and Hall (1994) the total potential for biomass in Europe is estimated to lie between 5 and 10 EJ, so that this scenario stretches the use of biofuels to their ultimate limits.

Table 12.11. Long-distance modal split in the scenarios (%)

Scenario / Mode	Market	Regulatory	Expected	Desired
Car				
ICEV conv.	0	0	30	0
ICEV alt. fuel	90	20	10	10
HEV	0	0	10	0
Fuel cell	0	0	0	10
Total car	90	20	50	20
Bus				
ICEV conv.	0	0	1	0
ICEV alt. fuel	5	0	1	5
HEV	0	0	3	0
Fuel cell	0	0	0	15
Train	0	60	27	40
HST	3	15	8	15
Maglev	1	5	0	0
Total coll. modes	9	80	40	
Plane				
Conventional	0	0	10	
LH_2	1	0	0	
Total	100	100	100	100

Despite the large mobility increase (82 per cent), the final energy consumption is comparable to that in 1989, due to the implementation of ICEVs which are significantly more energy efficient than they were in 1989. The total primary energy consumption in the market scenario is, however, larger than in 1989. This is due to the relatively poor conversion efficiency with which motor fuels can be derived from raw biomass (see Table 12.3). The consumption of fossil fuels and the consequent emission of CO_2 is, however, very low in this scenario (95 per cent below the level of 1989).

The regulatory scenario

The regulatory scenario combines a relatively slow increase of overall mobility

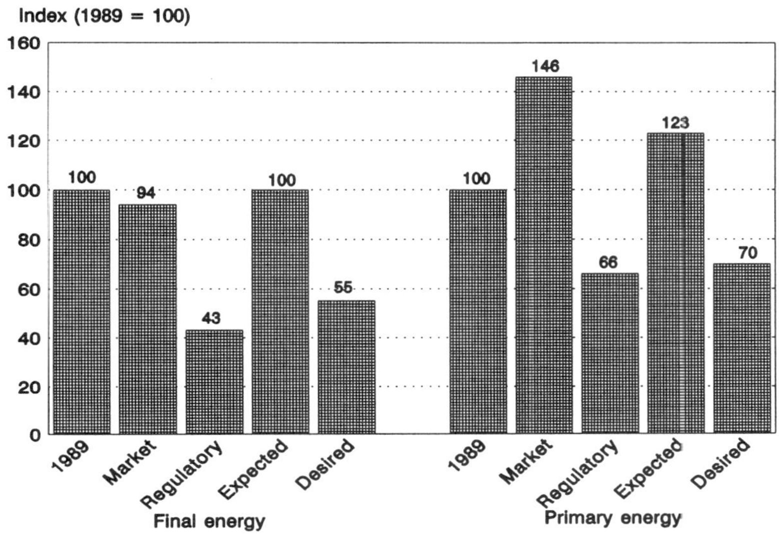

Figure 12.2. Use of final and primary energy in the scenarios

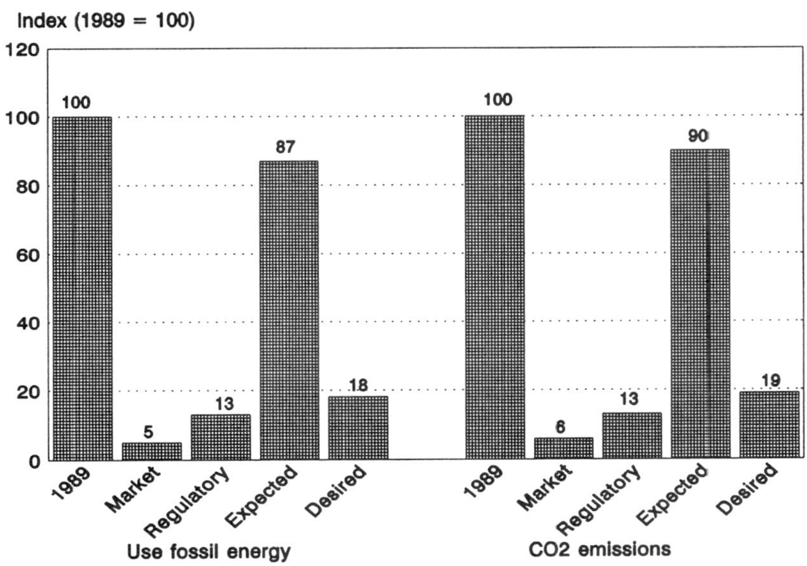

Figure 12.3. Use of fossil energy and the CO_2 emissions in the scenarios

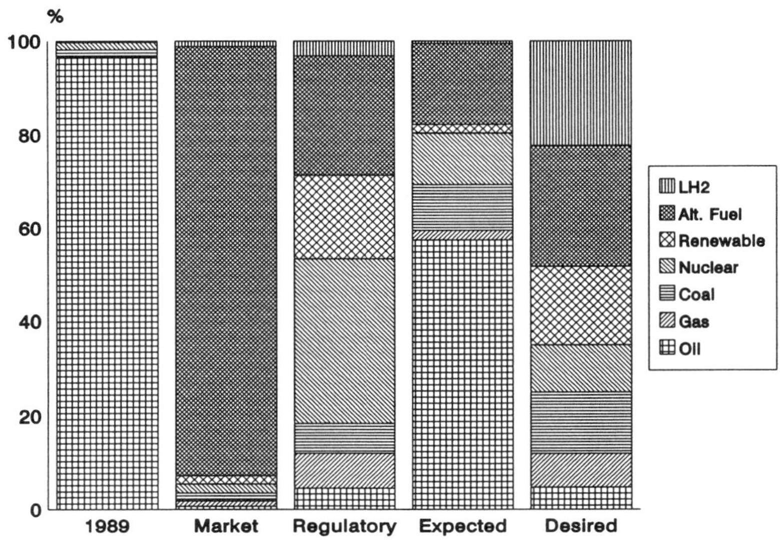

Figure 12.4. Shares of primary energy carriers in the scenarios

(46 per cent) with a dramatic shift in the modal split towards collective transport modes. In total 90 per cent of all urban passenger transport and 80 per cent of all long-distance transport is assumed to be performed by means of collective transport modes. The calculations show that for this scenario final energy demand can be reduced by more than 50 per cent, compared to 1989. The primary energy demand is reduced by only some 30 per cent. The shift to collective modes implies a shift to electricity as final energy carrier. In this scenario it has been assumed that a large part of the electricity is generated in nuclear power plants, which have a relatively low conversion efficiency resulting in a rather high primary energy consumption. As, however, electricity in the regulatory scenario is for the most part produced from non-fossil energy sources (nuclear and renewables), the resulting total fossil energy consumption and CO_2 emission is very low (about 88 per cent below the level of 1989).

The expected scenario

Mobility in the expected scenario is the highest of all scenarios (115 per cent above the level of 1989). Despite efficiency improvements in transport technologies the final energy demand in this scenario is comparable to that of 1989. Although much less than in 1989, the transport system is still dominated by individual transport by means of passenger cars (about 50 per cent of total mobility). For the most part these are still powered by fossil motor fuels, but there is also some introduction of biofuels and electricity. Collective transport relies on buses, which are mainly diesel powered. Fossil energy consumption and CO_2 emissions in the expected scenario are therefore high compared to the other scenarios, although slightly (10 per cent) lower than in 1989. Primary energy consumption is higher than in 1989, but lower than in the market scenario.

The desired scenario

To some extent the desired scenario is similar to the regulatory scenario. Both are characterized by a significant shift from individual to collective transport modes, although this shift is less drastic. Also the use of renewable energy sources for electricity production and biofuels for transport is comparable. An important difference, however, lies in the fact that the use of nuclear energy has been greatly reduced. In the desired scenario this is partly compensated by the extensive use of hydrogen, and partly by using electricity from conventional primary sources (coal and gas). With a figure of 77 per cent, the mobility increase is also much larger than in the regulatory scenario. All in all this still results in a significant decrease in final and primary energy consumption, compared to 1989. The fossil energy consumption and CO_2 emissions are slightly higher than in the market and regulatory scenario, but still 80 per cent lower than in 1989.

12.6 CONCLUSIONS

It appears that the market, regulatory and desired scenario *all* lead to a substantial reduction of the CO_2 emissions, at least as far as the European passenger transport sector is concerned. These reductions are even higher than the reductions required in the Toronto Declaration for the year 2025, which indicate an 80 per cent reduction of the CO_2 emissions (WMO, 1988; see also Section 2.3.2). In the expected scenario, however, only a stabilization at about the level of 1989 is achieved.

A general conclusion from the previous analyses may be that, provided the assumptions made are more or less correct, in the long term a sustainable transport system can, in principle, be realized by means of rather different sets of spatial, institutional, economic, socio-psychological and technological changes in our society and in its transport system. The required changes, however, need to be very drastic in all three scenarios that reach a sufficient level of reduction of CO_2, and in all cases the sustainability level can only be reached if transport technologies are employed which are much more energy efficient than present-day technology. Therefore, also drastic improvements of current transport technologies are necessary.

13 THE REALISM OF SUSTAINABLE MOBILITY: AN ASSESSMENT OF POLICY OPTIONS

13.1 INTRODUCTION

Is there still hope and scope for a sustainable transport system? Or are all the good intentions and policies bound to fail? From the scenarios presented in Chapters 8 to 10 and the calculations in Chapter 12, it may be concluded that the road towards an environmentally friendly transport system is likely to be a very hard one, but not an impossible one. This holds especially true for the reduction of CO_2 emissions, as this is directly related to energy use in the transport sector. In this final chapter we will analyse which policy options are possible and which choices have to be made to achieve a (more) sustainable transport system. Therefore, in the first section a general overview will be given of possible policy responses to the greenhouse effect, followed by a discussion of the implications for the transport sector arising from the scenarios developed in this study. Next, an assessment is given of the extent to which the variables on the various Spider axes are likely to influence the future of our transport systems and of the way policy makers may interfere in this sector. Finally, some conclusions related to policy choices will be offered.

13.2 POLICIES ON THE GREENHOUSE EFFECT

In the previous chapters various policy responses have been mentioned. Governments may indeed react to the climate change problem in many ways, by developing both short-term and long-term strategies. Policy making appears to be influenced by various background causes, uncertainties, institutional ramifications, international relationships, values and norms. In principle, at least five types of generic policy orientations may be distinguished (Klabbers *et al.*, 1994):

- no regrets;
- least regrets;
- acceleration;
- technological innovation;
- institutional/cultural change.

We will first briefly elaborate these five orientations, while in the next section we will apply them to the transport sector.

In the *no regrets* strategy, a minimum risk policy strategy is taken for granted, based on the fact that climate change is not yet entirely significantly proven. So it is questioned whether strict and large-scale reductions of greenhouse gas emissions are necessary. Priority may therefore be given to instruments which at least also serve other objectives (e.g. societal, such as employment) and which simultaneously result in a reduction of greenhouse gas emissions. In this way policies will pay off, even if the climate change does not happen as a global phenomenon.

A policy orientation with more far-reaching impacts is the *least regrets* option. The climate change problem is acknowledged here, but the impacts are considered to be still uncertain. Therefore, a trade-off is made between impacts of policies and the possible climate change effects. The policies are mainly anticipatory, in order to reduce possible impacts of climate change. Emissions are not reduced to the maximum extent, but when further reductions appear to be necessary, this can be achieved more easily -with less efforts - than in the no regrets option.

A policy choice aiming at an intense but generally sustainable development on a global scale is the *acceleration* option. In this strategy, sustainability is the main objective, while climate policies are an instrumental part of it. Not only greenhouse gas emissions, but also other sustainability criteria are given a central position in public policy. Therefore, a trade-off has to be made between a reduction of greenhouse gases and a reduction of other externalities. The measures proposed, however, take current trends and forces in society for granted.

Next, the *technological innovation* option recognizes that a (global) sustainability problem does exist, but expects that technological development will be sufficient for curbing the climate change problems. Policy is oriented at stimulating new technical developments (by subsidies, stimulating R & D, etc.) which may contribute to a sufficient reduction of greenhouse gas emissions and other externalities.

The final policy orientation is that of *institutional/cultural* change, which assumes that technical solutions *per se* will not be sufficient for achieving sustainability. Therefore, many aspects in the societal, institutional and cultural field have to change. The role of the government is largely limited to providing incentives, directions and support for behavioural change at the individual level.

It is clear that these distinct directions each have a different time horizon. No regrets policies may be introduced in the short term, while institutional/cultural changes may take at least 20 years. In this study we have mainly focused on long-term developments, in which a need for reduction of greenhouse gases is presumed. Therefore, in our study mainly technical innovations and institutional/cultural changes have been discussed. It is clear, however, that short-term strategies may also make a valuable contribution to reductions of greenhouse gas emissions.

In the next section we will explore the implications of all these remarks for policy options regarding the future of transport systems. We will conclude that the mission of sustainable mobility is certainly not impossible, but requires immediate and intensified efforts.

13.3 GENERAL POLICY OPTIONS IN TRANSPORT

13.3.1 Trends in transport policies

From the previous chapters it has become clear that the human geographical action space is structurally broadening. There seems to be a long-term megatrend from a low mobility society in the past (where only nomads, scientists, artists, soldiers, tradesmen and missionaries were geographically mobile) to a mass mobility society (in which not a select few but society at large is exhibiting a geographical drift). In contrast to the English expression 'my home is my castle' we observe nowadays a trend towards 'the globe is our home'. Transport is obviously the trademark of the mobile society.

At the same time it should be recognized that transport causes a wide array of externalities. These may be either positive (in the sense of positive pecuniary impacts of network operations as advocated by Scitovsky) or negative (in the sense of social costs as defined by Pigou). The assessment of all relevant social costs and benefits of the transport sector is fraught with many difficulties because of lack of data, lack of a clear measurement methodology or lack of a solid scientific background.

Policies to cope with such externalities cannot yet boast of a high degree of success. Regulatory systems have often failed because of too ambitious target levels, an insufficient use of instruments or lack of insight into behavioural responses. As a result, a broad spectrum of government failures in the transport sector has arisen, which has done nothing to contribute to the development of environmentally sustainable transport systems.

In this context, it is noteworthy that in recent years a marked shift in emphasis towards economic principles for a combined transport and environment policy can be observed. First, there is the trend towards abolishing unjustified and unnecessary protectionist or privileged regulations in order to increase the efficiency of transport operations and hence also to save the environment. Secondly, various user-charge principles are increasingly being discussed and implemented, such as road pricing, toll principles, parking fees and perhaps in the long run even tradeable permits. Such measures may serve to keep the contribution of the transport system to environmental pollution below a sustainable threshold of environmental decay.

Against this background attention to distributional issues is also in order.

Transport systems have different degrees of access for different modes and different users, and at the same time generate unequal distributions of environmental decay. In this respect, tradeable permits for mobility (e.g. at the urban level) are becoming an increasingly popular tool (see also Vleugel, 1995).

Thus it would seem that economic principles are gaining more importance in future transport policies, although it should be recognized that economic optimality is not automatically ensured, unless such economic principles are applied within a strict context of an *a priori* determined 'environmental utilization space' for transport (see Section 2.5). It is clear that we encounter here a rich research field.

Several of the above observations were directly or indirectly also addressed in our scenario experiments. In Section 13.2 we made a typology of five generic policy orientations. Policies - aiming at sustainability in the transport sector - may also be divided in the same way for each of the scenarios distinguished. We will deal with this issue in the next section.

13.3.2 An assessment of transport policy options

In general, policies focused on making transport systems more sustainable may use four practical measures to reduce greenhouse gas emissions by the transport sector, viz.:

- an improvement of cars and planes;
- the introduction of new technologies;
- a modal shift towards cleaner modes;
- a reduction of the total mobility level.

Each of these policy possibilities contributes to a rise in environmental sustainability. Clearly, in practice, policies may be a mixture of these four options. The main question is: What would the composition of this mixture have to be? This depends largely on the policy of the government, whose orientation may tend towards one of the five policy options discussed earlier.

No regrets policy

In this policy option the emphasis is put on achieving at least other goals as well. The improvement of cars and planes is an option, because this may increase energy efficiency and therefore lower production and transport costs, which has positive impacts on trade and economic growth. Moreover, an added bonus is the possibility that the dependence on oil-producing countries may be reduced. A modal shift towards collective modes may be attractive in order to reduce congestion, and therefore this may be a goal of government policy, too. The same holds for the introduction of some new technologies, like telematics and the HST. These will only be introduced, however, when the costs are considered to be acceptable.

Least regrets policies

In this option the 'no regrets' policies may be introduced even more extensively. Investments in collective modes may increase, which make a modal shift possible. Also R & D of new technologies may be stimulated more than in the no regrets option, so that these can be introduced when they appear to be necessary. In this way, society and individuals are prepared for possible changes in the future.

Acceleration policy

Here, a mixed policy is created from all four options, and other externalities are also taken into account to produce the most favourable mix. The policies try to incorporate current trends in transport and do not rely only on drastic behavioural changes or technological breakthroughs. Such a strategy is particularly relevant if the future situation is to resemble the expected scenario. It is clear, however, that it will not be easy to make such a policy an unqualified success. Governments in democratic countries will tend to introduce policy measures which cause the least resistance. In general, the improvement of current modes will cause less antagonism in society, because individual behaviour and the organization of society do not have to change too much. As discussed in Section 5.2, the expectations are that between 1990 and 2000 the fuel consumption of private cars may reduce by 15 to 22 per cent per kilometre, and after this period a further efficiency increase of some 25 per cent may be expected. It may be assumed, however, that mobility will grow even faster, and the same may hold for the aviation system. Therefore, it is clear that this - seemingly appealing - option is just not sufficient to achieve a sustainable transport system. But it is obvious that substantial reductions in mobility which would also be necessary often meet with much resistance, so that here a conflict between sustainability and social acceptability may arise.

Technological innovation

Clearly, the introduction of new technologies may cause the least resistance. Technical developments of individual modes and of high-quality collective modes may largely contribute to reductions of greenhouse gas emissions. In this framework a shift may also occur to more environmentally friendly transport modes (mainly collective). This, however, affects individual freedom and makes societal changes necessary. In general though, most efforts are being concentrated on the development of technical improvements of the car and plane and especially on the introduction of new technologies. A government may stimulate these developments by granting subsidies, introducing regulations, raising the price of conventional fuels, etc. Other options are not widely introduced, because the technological options are considered to be sufficient. The scenario best fitting this option may be the market-oriented reference scenario.

Institutional/cultural change

Technological innovations are - in this policy orientation - not thought to be sufficient for reducing greenhouse gas emissions. Therefore, a policy on institutional/cultural change is mainly aiming at a modal shift and mobility reduction. Individual freedom may be reduced in this way, and many other changes in society will also be necessary. The scenario supporting this option may be either the regulatory or the desired scenario. Of the two the latter will need less drastic changes in individual and societal behaviour.

13.4 POLICY OPTIONS AND CHOICES RESULTING FROM THE SCENARIOS

From the scenarios developed in this study, it may be concluded that changing the transport system in order to make it environmentally more sustainable has significant impacts in many fields. The lifestyle of individuals may be affected, for example, by intervening not only in their personal choice of where to live, but also in their freedom to choose their way of travelling. It should be understood, however, that the existing transport system is in fact very much the result of a whole history of developments in other fields. Changes in the emphasis on equity or a reversal of the individualization trend, for example, will be largely independent of purely transport motives, as is shown by past developments. Governments in democratic countries will usually follow these trends and hence may show no (major) intentions to intervene (see also Rietveld, 1997). In this context cultural aspects do play a definite role. In the United States, for example, intervention by the government tends to face more public resistance than in most European countries.

Given the need for a structural reduction of the contribution of transport to global environmental problems, it may be clear that several choices have to be made about the future of our transport system and its use. In our study these choices have been presented in an extreme way in two reference scenarios. For the current transport system the basic choice concerns either improving the existing transport system, which is mainly based on individual modes of transport (by introducing new fuels, increasing fuel efficiency, etc.), or introducing a more or less collective transport system in order to replace (part of) the current transport system.

As might be anticipated, the scenarios based on expert opinions are not that extreme, but reflect more realism. In our opinion, however, especially the *expert-based expected* scenario is rather too conservative. In fact, this scenario is largely an extrapolation of current trends. So it is questionable whether the realization of this scenario will lead to a sufficient reduction of negative environmental externalities. It appears, for example, that new technologies are not expected to play a revolutionary role in the future, while public policy is not expected to introduce the necessary policy measures for reaching environmental targets.

In the *expert-based desired* scenario the rather traditional choice is made that collective modes should dominate the transport scene, while the private car is only to be used as a feeder system for this collective system. According to most experts, government policy should aim at achieving such a collective transport system. A striking finding is that most experts - irrespective of age, gender, discipline or professional background - largely agree on both scenarios; the variance in answers between the subgroups was smaller than was expected.

Our statistical analysis of the CO_2 emissions of the various scenarios has shown, however, that we do not face an entirely hopeless situation. A sustainable transport system may become a reality. One interesting question may be where government policies could act to achieve more sustainability and indeed whether such intervention is desired. Therefore, we will elaborate on this further in the next sections by referring again to the four quadrants of the Spider model.

13.4.1 Spatial factors

European level

At the European level market forces may lead to a continuing spread of activities and working places. In theory, this is favourable for an individual transport system, which may be environmentally sustainable only if conventional fuels are replaced by alternatives like hydrogen and zero-emission electricity. If, however, a shift is desired towards collective modes, the specialization and concentration model is favourable in theory. Nevertheless, when the expert scenarios are analysed, it appears that in the desired scenario a more diffuse spatial organization is combined with a more collective transport system than in the expected scenario. An explanation of these results may be that a completely concentrated spatial organization has many negative consequences in other areas, for example, living conditions, noise annoyance, visual intrusion, congestion, etc.

Urban level

Just as concentration favours collective modes at a European scale level, this is also the case for the compact city concept at the urban level (ECMT/ OECD, 1995). Even so from the expected scenario it appears that even in a quite compact city individual modes may still be dominant, but it is nonetheless clear that a compact structure is favourable for collective modes. However, the living conditions in a compact city may be less favourable. Therefore, a trade-off has to be made in transport - and in general spatial - policy between living conditions (and other societal impacts) and wider environmental consequences.

Implications for spatial policy
The close relationship between the spatial organization and the transport system is widely acknowledged. Spatial policy is therefore of major importance for the future of transport. Other factors also, like living conditions (big vs. small dwellings; large vs. small household sizes; green suburbs vs. high buildings) and regional development play a major role in this policy field. Societal trends like individualization also impact on the preferred living conditions.

A concentrated spatial development may be an important success factor for a shift towards collective modes, although in the desired scenario this option is combined with a moderate diffuse spatial organization. The separation of responsibilities for planning and investment, however, is likely to hamper this development (ECMT/OECD, 1995), but it may only be possible to achieve such a shift by considerably increasing the real costs of car use.

It may be concluded, however, that when the policy path chosen is towards a collective transport system, government intervention encouraging a more concentrated and compact spatial organization (strict spatial planning) seems necessary. Important, too, is the fact that the planning of the transport system is a major variable in the policy decisions concerning the development of new living quarters and working areas.

13.4.2 Institutional factors

Regulation vs. market-based intervention
If government intervention is assumed to be necessary for achieving a more sustainable transport system, the question is: How should this be shaped? Two extreme policies are possible: regulation vs. market-based intervention. When introducing these two options, a trade-off has to be made between equity and efficiency, which also has impacts on other fields and is influenced by societal trends at large.

Market-based intervention may lead to a mainly individual transport system accompanied by the introduction of new fuel cars, which is the best case from the point of view of alleviating the greenhouse effect. In these circumstances, mobility has to become a luxury good, since otherwise society would have to sacrifice too many scarce resources for the transport system. The latter situation may lead to a more collective transport system. This relationship is largely confirmed by the expert scenarios developed in our study.

Management of modes and infrastructure
In this framework, the important question is how to organize the management of modes and infrastructure. When this is the responsibility of public companies, the emphasis may be placed on equity, while efficiency is low, although for public companies efficiency incentives may also be introduced. Moreover, infrastructure may be publicly or privately owned. It may be assumed that public management is in favour of collective modes, while privatized transport may lead to a more individual system according to the scenarios.

Implications for institutional policy
A trade-off has to be made between efficiency and equity, which may lead to an entirely different transport system. In the past decade, there has been a clear trend towards privatization and deregulation in the transport sector. But these policies may change over time, because views on the role of the government may also change as has happened in the past. However, a stable institutional environment is important, because then uncertainty is reduced, which may stimulate investments in sustainable transport options.

13.4.3 Economic factors

Subsidized vs. profitable transport
Official policies often aim at internalizing external costs in the transport sector, which may lead to price increases affecting the private car in particular, but also collective modes. In practice, however, there is much resistance in society against these policies, because individual freedom (by cheap mobility) is hampered. However, if it is decided to reduce car (and plane) use, price increases especially of conventional fuels may lead to the introduction of new fuels, a modal shift towards collective modes and a reduction of car mobility.

On the other hand, price increases of collective modes because of cuts in subsidies may lead to a closing down of network links and to lower frequencies, which may reduce the use of collective modes.

Market principles vs. coordination
Coordination and centralization at a European level may lead to a level playing field in transport, so that no competitive advantages may be achieved. In this case there may be a major role for the European Union. For example, international cooperation of railway companies may be improved by European coordination, which may be a very important success factor for, say, high-speed trains. At the

same time, however, competition between transport operators or integrators would have to be increased. On the other hand, such a centralization may hamper economic growth, which may again reduce the available funds for investments in new technologies.

An opposite development may be that increasing national or regional competition emerges, which may stimulate economic growth. Measures may differ per region, though these may not be introduced in some regions for economic reasons (free-rider behaviour). It may be expected that such a decentralized economy favours individual modes, especially on long distances for which collective modes have the competitive edge.

Implications for economic policy

In economic policy a choice has to be made between coordination/centralization and deregulation, which has an enormous impact on transport. However, other objectives such as economic growth, the social security system, societal views, etc. may be more important for such a decision than transport. Society has also to decide on the profitability of transport and on the extent to which it wants to set aside scarce funds for transport and for the mobility possibilities of individuals.

It is clear that the current devolution trends may lead to a more market-oriented transport system, which is mainly individually oriented.

13.4.4 Socio-psychological factors

Equity vs. inequity

The emphasis on equity largely depends on other factors, like the way the economy and the transport market are organized. In society itself, however, strong forces may also influence the political emphasis on equity. In the United States, for example, this issue is less important than in Western-European countries where large-scale social security systems have been introduced.

For the transport market an emphasis on equity may lead to a subsidized collective transport system, where car use is restricted. Equity would be achieved, because then mobility would be available for every individual.

Individualization vs. social cohesion

A continuing individualization is a definite trend in society, which may be influenced by public policies only to a small extent. However, this trend may be reversed in the long run, as is assumed in the cultural/institutional change option discussed above. Individualization may lead to more market-oriented measures and a diffuse spatial organization, because of the growing amount of small households and a continuing emphasis on individual modes. Social cohesion, on the other hand, may stimulate regulation and the use of collective modes.

Implications for the socio-psychological field
Governments may try to steer social developments, but it is questionable whether a big impact of these policies may be expected. Governments in democratic countries are dependent on their citizens and may therefore largely follow these trends instead of trying to curb them. It is evident, however, that the societal view on many aspects is very important for public policy in other force fields of policy.

13.5 REALISM OF SUSTAINABLE MOBILITY

The road towards a sustainable transport system is paved with good intentions, but is facing major stumbling blocks due to the fact that transport decisions are based on derived individual and societal objectives. Changing the transport system appears to have far-reaching consequences for many aspects of society and individual behaviour, but it should be recognized that many other social, cultural political and economic factors also play a role. For a sustainable future of transport it appears to be important to make several strategic policy choices.

The *first policy choice* is to decide whether one favours a transition towards a stronger stimulation of *collective* modes, or whether one gives priority to the development of cleaner *individual* modes. It is not efficient to develop both systems simultaneously, because of the high transaction costs and the different developments needed in many fields. From the scenarios it appeared that extensive trade-offs would need to be made. Each of these compromises has profound impacts in other fields, which are also important for the welfare of individuals and society at large. There is apparently a yawning gap between wish and reality.

Essential for this choice - as a *second policy choice* - is the future *spatial organization*. If a spatial organization is achieved that is in agreement with the concentration and specialization scenario and compact city concept, then collective modes have a better potential. Otherwise, individual modes will tend to dominate the scene. In practice, however, a trade-off has to be made between the environmental, societal and other external impacts. What is clear is that the spatial organization is one of the variables which can be influenced by public policy.

An important conclusion from our study is that new transport technology options are in general not competitive with existing modes, the car and the plane, because of the high quality and relatively low price of these latter options.

A *third policy choice* may therefore be to deliberately make the conventional *car and plane* unattractive despite resistance in society. Two policy directions are possible in this case, which are also dependent on developments in the socio-psychological field.

First, *market* measures may be introduced. One may think of raising (and for air traffic introducing) taxes on conventional fuels up to a level of environmentally-friendly behaviour. The introduction and application of alternative fuels may be greatly stimulated in this way, while the competitiveness of collective modes may then be increased. This last goal can also be achieved by the introduction of road- and cordon-pricing systems. An additional advantage of these measures may be that overall mobility will then become much more expensive, which reduces mobility growth. Such measures may be compatible with the extreme market-oriented reference scenario.

In the second place it is possible to apply stronger *regulation*. One can think of more severe environmental targets, which may make it more attractive to use alternative fuels. It is also possible to introduce compulsory emission targets for specific modes, as is the case for the zero-emission car in California. In urban areas it is also possible to reduce the density of the road network and to introduce a strict parking policy. In this way the use of collective modes may be stimulated, as in the regulatory and desired scenario.

A *fourth policy choice* is that the government may stimulate the introduction of new options by means of *offering convincing examples* of both good intensions and actions. The government may do this by purchasing zero-emission cars for several public organizations despite the provisionally high costs. In this way the development of cleaner technologies may be stimulated, because it is guaranteed in this case that development costs will be paid back. As a result, new transport technologies become cheaper for other parties, which, in turn, will stimulate their adoption. Another possibility is to encourage the use of collective modes by public employees, e.g. by offering them discount on public transport fares.

Our study has convincingly demonstrated that 'business as usual' is an unviable strategy to attain a sustainable future for the transport sector. From the scenarios in our study, it may be concluded that in the *expected* future environmental goals are not likely to be attainable at all in the year 2030, while in the *desired* situation much more intensified government action and subsequent changes in individual behaviour and the transport technology used are necessary. It is a common view among transportation experts, that the desired option is not in easy reach, since too many factors have to change within society. On the other hand, the time horizon is still sufficiently far away, so that it is not too late to induce substantial changes in our transport system. Drastic shifts in the economic, institutional and social environment occur more often and quicker nowadays than in the past; a situation which may continue in the future. So a shift from the 'expected' to the 'desired' future may be a feasible option. In any case, a sustainable transport system is in principle a feasible outcome if the desired option is pursued.

In conclusion, based on our application of the Spider scenario model, it appears that several roads may be travelled towards an environmentally more friendly transport system, but whatever the road ultimately chosen, it will be a hard one to follow. Without strict policy measures, for the next generation environmentally sustainable mobility will probably still seem more of a challenge than a realistic objective. Only far-reaching changes in lifestyle, technology and policy will be able to turn the tide.

REFERENCES

Adriaanse, A. (1993) *Environmental Policy Performance Indicators*, SDU, The Hague.

Amin, A. (1993) 'The Globalization of the Economy: an Erosion of Regional Networks?', in Grabher, G. (ed.), *The Embedded Firm. On the Socio-Economics of Industrial Networks*, pp. 278-295, Routledge, London.

Andersen, B. (1993) 'A survey of the Swiss Public Transport System and Policy', *Transport Reviews*, vol. 13, no. 1, pp. 61-81.

Banister, D. and S. Watson (1994) *Energy Use in Transport and City Structure*, Planning and Development Research Centre, University College, London.

Barde, J.-P. and K.J. Button (eds.) (1990) *Transport Policy and the Environment*, Earthscan, London.

Batenburg, R.S. and W.P. Knulst (1993) *Sociaal-Culturele Beweegredenen. Onderzoek naar de Invloed van Veranderende Leefpatronen op de Mobiliteitsgroei sinds de Jaren Zeventig*, Sociaal Cultureel Planbureau, Rijswijk.

Berg, J.R. van den (1993) *Nieuwe Vervoerssystemen en Technieken*, Stichting Postacademisch Onderwijs in de Vervoerswetenschappen en de Verkeerskunde, Delft.

Bergh, J.C.J.M van den and P. Nijkamp (1997) 'Global Environmental Change, Local Land Use Impacts and Socio-Economic Response Strategies in Coastal Regions', *Journal of Coastal Zone Management* (forthcoming).

Biéber, A., M.H. Massot and J.P. Orfeuil (1994) 'Prospects for Daily Urban Mobility', *Transport Reviews*, vol. 14, no. 4, pp. 321-339.

Bilderbeek, R.H., W. Korver and J. Schot (1993) *Technische Innovaties in het Personenverkeer en -vervoer: Een Inventarisatie op Zoek naar Duurzame Mobiliteit*, STB-TNO report no. 93/28, Apel doorn.

Boer, E. de (1989) 'Vervoermiddelbehoefte en Vervoersmogelijkheden', in Knippenberg, C.W.F. van, J.A. Rothengatter and J.A. Michon (eds.), *Handboek Sociale Verkeerskunde*, Van Gorcum, Assen.

Boonekamp, P.G.M., O. van Hilten, P. Kroon and M. Rouw (1992) *Nationale Energie Verkenningen 1990-2015*, ECN-C-92-017, Energy Centre of the Netherlands (ECN) Policy Studies, Petten.

Bovy, Ph. (1994) 'Neun Fragen zu Swissmetro', *Polyrama*, February, pp. 23-25.

Bovy, P.H.L., J.P. Orfeuil and D. Zumkeller (1993) 'Europe: a Heterogeneous

Single Market', in Salomon, I., P.H.L. Bovy and J.P. Orfeuil, *A Billion Trips a Day*, Kluwer, Dordrecht.

Brand, M. (1987) The Competitive Niche of Maglev Compared to High Speed Rail and Air, paper presented at the 13th International Conference on Maglev and Linear Drives, May 19-21, Las Vegas.

Breheny, M. (1994) 'Counterurbanisation and Sustainable Urban Forms', in Brotchie, J., M. Batty, P. Hall and P. Newton (eds.), *Cities in Competition: The Emergence of Productive and Sustainable Cities for the 21st Century*, Longman Cheshire, Melbourne.

Breimeier, R. (1993) 'Die Einsatzmöglichkeiten der Magnetschwebebahn Transrapid: Weiterer Diskussionsbeitrag', *Internationales Verkehrswesen*, vol. 45, no. 4, pp. 184-190.

Brotchie, J.F., M. Anderson and C. McNamara (1994) 'Changing Metropolitan Commuting Patterns', in Brotchie, J.F., M. Batty, P. Hall and P. Newton (eds.), *Cities in Competition: The Emergence of Productive and Sustainable Cities for the 21st Century*, Longman Cheshire, Melbourne.

Bruinsma, F.R. (1994) *De Invloed van Transportinfrastructuur op Ruimtelijke Patronen van Economische Activiteiten*, PhD thesis, FEWEC, Free University, Amsterdam.

BTS (Bureau of Transportation Statistics) (1996) *Transportation Statistics, Annual Report 1996*, Dept. of Transportation, Washington DC.

Bunch, D.S., M. Bradley, T.F. Golob, R. Kitamura and G.P. Occhiuzzo (1993) 'Demand for Clean-Fuel Vehicles in California: A Discrete-Choice and Stated Preference Pilot Project', *Transportation Research*, vol. 27A, no. 3, pp. 237-253.

Bureau Ferro (1991) *Uitkomsten van een Onderzoek naar de Betekenisstructuur van het Niet-Zakelijk Autogebruik volgens de Ladderingmethode*, Research Report, Amsterdam.

Büttner, T. and A. Grübler (1994) *The Birth of a 'Green' Generation? Generational Dynamics of Resource Consumption Patterns*, International Institute for Applied Systems Analysis (IIASA).

Button, K. (1993) 'The Future of European Transport' in Thord, R. (ed.), *The Future of Transportation and Communication*, Springer-Verlag, Berlin.

Button, K. and D. Pitfield (eds.) (1991) *Transport Deregulation: An International Movement*, Macmillan, Basingstoke.

Carlson, R.C., W.W. Harmen, P. Schwartz and associates (1982) *Energy Futures, Human Values and Life Styles. A New Look at the Energy Crises*, Westview Press, Boulder, Co.

CBS (1992) *Statistiek van het Personenvervoer*, CBS nr. N15, Central Bureau of Statistics, Voorburg/Heerlen.

CBS (1994) *De Mobiliteit van de Nederlandse Bevolking 1993*, CBS nr. N8, Central Bureau of Statistics, Voorburg/Heerlen.

Cedrún, G. and G. de Rus (1994) 'Spain's Transport Policy', *Journal of Transport Economics and Policy*, vol. 28, nr. 2, pp. 211-214.

Clerx, W.C.G. and E.J. Verroen (1992) *Ruimtelijke Inrichtingsvarianten voor Nederland: Vervoerspatronen en Milieuconsequenties*, INRO-TNO, Delft.

Commission of the EC (1993a) *The Future Development of the Common Transport Policy, a Broad Approach for the Establishment of a Community Framework for Sustainable Mobility*, Brussels/Luxembourg.

Commission of the EC (1993b) *European High Speed Rail Network: A Socio-Economic Impact Study*, Luxembourg/Brussels.

Commission of the EC (European Communities) (1993c) *Trans-European Networks: Towards a Master Plan for the Road Network and Road Traffic*, Motorway Working Group, Brussels.

CPB (1992a) *Scanning the Future*, Central Planning Bureau, The Hague.

CPB (1992b) *Verkeer en Vervoer in Drie Scenarios*, working document 35, Central Planning Bureau, The Hague.

Datar/Reclus (1989) *Les Villes < <Européennes> >*, Paris.

Davelaar, E.J. and P. Nijkamp (1992) 'Operational Models on Industrial Innovation and Spatial Development: A Case Study for the Netherlands', *Journal of Scientific & Industrial Research*, vol. 51, pp. 273-284.

DeLuchi, M.A. (1993) 'Greenhouse-Gas Emissions from the Use of New Fuels for Transportation and Electricity', *Transportation Research*, vol. 27A, nr. 3, pp. 207-216.

DeLuchi, M.A. and J.M. Ogden (1993) 'Solar-Hydrogen Fuel-Cell Vehicles', *Transportation Research*, vol. 27A, nr. 3, pp. 255-275.

Dimitrou, H.T. (ed.) (1994) *Moving Away from the Motor Vehicle*, Centre of Urban Planning and Environment Management, Hong Kong.

Dreborg, K.H. and P. Steen (1994) A Swedish Transportation Futures Study, Research Paper of the Environmental Strategies Research Group, Stockholm.

Duenk, F.H.J. (1991) *Het Infrastructuurbeleid van de Europese Gemeenschap*, Infrastructure, Transport and Logistics Series, nr. 8, Delftse Universitaire Pers, Delft.

ECMT (1990) *Private and Public Investment in Transport*, European Conference of Ministers of Transport, Paris.

ECMT (1992) *Guided Transport in 2040*, European Conference of Ministers of Transport, Paris.

ECMT (1993) *Annual Trends in Transport 1965-1989*, European Conference of Ministers of Transport, Paris.

ECMT/OECD (1995) *Urban Travel and Sustainable Development*, European Conference of Ministers of Transport, Paris.

Ellwanger, G. and M. Wilckens (1993) Hochgeschwindigkeitsverkehr Gewinnt an Fahrt, *Internationales Verkehrswesen*, vol. 45, nr. 5, pp. 284-290.

Emanuel, J.D. (1991) Ultra Light Mass Transit; Today's Costs and

Environmental Breakthrough in Mass Transit and People Movers, research paper.
Emmerink, R.H.M. (1996) *Information and Pricing in Road Transport*, Ph.D. Thesis, Free University, Thesis Publishers, Amsterdam.
Emmerink, R.H.M., K.W. Axhausen, P. Nijkamp and P. Rietveld (1994) 'The Economics of Motorist Information Systems Revisited' *Transport Reviews*, vol. 14, nr. 4, pp. 363-388.
Enquete-Kommission 'Schutz der Erdatmosphäre' des Deutschen Bundestages (Hrsg.) (1994) *Mobilität und Klima*, Economica Verlag, Bonn.
Eurostat, several years, *Basisstatistieken van de Gemeenschap*, Luxembourg/Brussels.
Eurostat (1993) *Demographic Statistics 1993*, CEE, Luxembourg.
Festinger, L. (1954) 'A Theory of Social Comparison Processes', *Human Relations*, vol. 7, pp. 117-140.
Festinger, L. (1957) *A Theory of Cognitive Dissonance*, Peterson, Evanston, ILO.
Fishbein, M. and I. Ajzen (1975) *Belief, Attitude, Intention, and Behavior: An Introduction to Theory and Research*, Addison-Wesley, Reading, MAO.
Fokkema, T. and P. Nijkamp (1994) 'The Changing Role of Governments: The End of Planning History?', *International Journal of Transport Economics*, vol. 21, nr. 2, pp. 127-145.
Frey, B.S. (1983) *Democratic Economic Policy: A Theoretical Introduction*, Blackwell, Oxford.
Fujino, M. (1993) Outline of HSST-100 System and Test Line in Nagoya, *Proceedings of the 13th International Conference on Magnetically Levitated Systems and Linear Drives*, May 19-21, Argonne Ill.
Gärling, T. (1993) Environmental Deterioration Caused by Traffic, paper presented at the Dutch Transport Planning Colloquium, (mimeographed), November, Rotterdam.
Garett, M. and M. Wachs (1996) *Transportation Planning on Trial*, Sage, London.
Garrison, W.L. (1989) 'Technology and Transportation Systems' in Batten, D.F., and R. Thord (eds.), *Transportation for the Future*, Springer-Verlag, Berlin.
Geenhuizen, M. van and P. Nijkamp (1995) 'Industrial Dynamics, Company Life Histories and the Core-Periphery Dilemma', *Geography Research Forum* (forthcoming).
Geerlings, H. (1995) The Rise and Fall of New Technologies; the Maglev as Technological Substitution?, *Proceedings TRAIL-congress*, 30-5-1995, Rotterdam.
Goddard, H.C. (1997) 'Sustainability, Tradeable Permits and the World's Largest Cities; A new Proposal for Controlling Vehicle Emissions, Congestion and Urban Decentralization with an Application to Mexico City', *Techology in Pollution*, vol. 7, nr. 3, pp. 357-374.

Grieb, H. and B. Simon (1990) 'Pollutant Emissions of Existing and Future Engines for Commercial Aircraft', in Schuman, U. (ed.): *Air Traffic and the Environment; Background, Tendencies and Potential Atmospheric Effects*, Springer-Verlag, Berlin.

Grontmij Advies and Techniek (1994) *Op Weg naar Verplaatsen binnen de Milieugebruiksruimte (2040)*, behoeftenveldanalyse in het kader van het Interdepartementaal Onderzoeksprogramma Duurzame Technologische Ontwikkeling (DTO), Zeist.

Grübler, A. (1990) *The Rise and Fall of Infrastructures: Dynamics of Evolution and Technological Change in Transport*, Physica Verlag, Heidelberg.

Grübler, A. (1993) 'The Transportation Sector: Growing Demand and Emmissions', *Pacific and Asian Journal of Energy*, vol. 3, nr. 2, pp. 179-199.

Grübler, A. (1994) 'The Growing Transport Sector: Growing Demand and Emissions', *Pacific and Asian Journal of Energy*, vol. 3, nr. 2, pp. 179-199.

Grübler, A. and N. Nakićenović (1991a) *Evolution of Transport Systems: Past and Future*, IIASA-report RR-91-8, Laxenburg.

Grübler, A. and N. Nakićenović (1991b) *Long Waves, Technology Diffusion, and Substitution*, IIASA-report RR-91-17, Laxenburg.

Gunn, H.F., M.A. Bradley and D.A. Hensher (1992) High Speed Rail Market, *IIASA-report RR-91-17*, Laxenburg, International Institute for Applied Systems Analysis (IIASA), Austria.

Gwilliam, K.M. and H. Geerlings (1994) 'New Technologies and their Potential to Reduce the Environmental Impact of Transportation', *Transportation Research*, vol. 28A, nr. 4, pp. 307-319.

Hall, D.O. (1994) Biomass Energy Options, paper presented at the International Energy Agency (IEA) conference on 'Energy Technologies to Reduce CO_2 Emissions in Europe: Prospects, Competition, Synergy', 11-12 April, Petten.

Hall, P. and D. Banister (1995) 'The Second Railway Age', *Built Environment* (forthcoming).

Hart, T. (1994) 'Transport Choices and Sustainability: A Review of Changing Trends and Policies', *Urban Studies*, vol. 31, nr. 4/5, pp. 705-727.

Heilig, G.K. (1994) 'The Greenhouse Gas Methane (CH_4); Sources and Sinks, the Impact of Population Growth, Possible Interventions', *Population and Environment*, vol. 16, nr. 2, pp. 109-137.

Hey, C. (1996) The Incorporation of the Environmental Dimension into the Transport Policies in the EU, Discussion paper nr. 50, Eures, Freiburg.

Hilten, O. van, J.M. Bais and A.D. Kant (1992) *Electric Vehicles in Urban Areas: A Case-Study for Amsterdam*, ECN-C-92-084, Energy Centre of the Netherlands (ECN) Policy Studies, Petten.

Himanen, V., P. Nijkamp and J. Padjen (1993) 'Transport Mobility, Spatial Accessibility and Environmental Sustainability', in Nijkamp, P. (ed.); *Europe on the Move*, Avebury, Aldershot, pp. 269-290.

Hondius, H. (1990) 'Met 250 km/h en meer door Europa', *De Ingenieur*, nr. 9, pp. 28-33.

Hoogenstrijd, G.J. (1993) *Luchtvaart Grenzeloos?*, IVEM-report nr. 71, State University Groningen.

Hoorn, I.J.M. van der (1989) 'Verplaatsingsgedrag als Afgeleide van het Activiteitenpatroon' in Knippenberg, C.W.F. van, J.A. Rothengatter and J.A. Michon (eds.), *Handboek Sociale Verkeerskunde*, Van Gorcum, Assen/Maastricht.

Houghton, J.T., G.J. Jenkins and J.J. Ephraums (eds.) (1990) *Climate Change, the IPCC Scientific Assessment*, Cambridge University Press, Cambridge.

Houghton, J.T., B.A. Callander and S.K. Varney (eds.) (1992) *Climate Change, the Supplementary Report to the IPCC Scientific Assessment*, Cambridge University Press, Cambridge.

Hughes, P. (1993) *Personal Transport and the Greenhouse Effect*, Earthscan, London.

IEA/OECD (1993) *Cars and Climate Change*, IEA/OECD Energy and Environment Series, International Energy Agency/Organisation for Economic Cooperation and Development, Paris.

Ingenieurskrant 20 (1990) 'Boren van Tunnels nog Vrij Kostbaar', September 20, p. 13.

Ingenieurskrant 23 (1993) 'TGV en ICE Zijn Tegenpolen', October 28, p. 12.

Intermediair (1990) 'De Zweeftrein Gaat Rijden', vol. 26, 05-10-1990, p. 39.

Intermediair (1993) 'De Zeppelin van de Volgende Eeuw', vol. 29, nr. 23, p. 37.

IPCC (1996) *Second Assessment Report*, International Panel on Climate Change, New York.

ISDS (1988) *Integraal Transport Systeem*, Rijkswaterstaat, Dienst Verkeerskunde, International System Development and Support, Delft.

ISDS (1989a) *Europe in Motion; the Potential of ITS - the Integrated Transport System*, International System Development and Support, Delft.

ISDS (1989b) *Masterplan ITS*, International System Development and Support, Delft.

ISDS (1989c) *Integraal Transport Systeem*, International System Development and Support, Delft.

ISDS (1993) *Betuwelijn Ondergronds*, Delft.

Jansweijer, R.M.A. (1987) *Private Leefvormen, Publieke Gevolgen, naar een Overheidsbeleid met betrekking tot Individualisering*, voorstudies en achtergronden, WRR, V57, SDU, The Hague.

Jeffery, D. (1990) 'Driver Route Guidance Systems: State-of-the-Art' in Soekkha, H.M., P.H.L. Bovy, P. Drewe and G.R.M. Jansen (eds.), *Telematics - Transportation and Spatial Development*, pp. 151-172, VSP, The Hague.

Jones, P. (1991) 'Gaining Public Support for Road Pricing through a Package Approach', *Traffic Engineering and Control*, April, pp. 194-196.

Kahn, A.E. (1966) 'The Tyranny of Small Decisions', *Kyklos*, vol. 19, pp. 23-45.

Kahn, H. and Wiener, A.J. (1967) *The Year 2000; A Framework for Speculation on the Next Thirty-Three Years*, Beckmans, Stockholm.

Kaminishi, K., A. Seki, and H. Tsuruga (1993) The Development of the Superconducting Maglev System, *Proceedings of the 13th International Conference on Magnetically Levitated Systems and Linear Drives*, May 19-21, pp. 10-15, Argonne Ill.

Kemp, R. (1994) *Technology and the Transition to a Sustainable Economy, Continuity and Change in Complex Technological Systems*, Paris.

Klaassen, G.A.J. (1995) *Trading Sulfur Emission Reduction Commitments in Europe*, Ph.D. thesis, Free University, Amsterdam.

Klaassen, G.A.J. and J.B. Opschoor (1992) 'The Economics of Sustainability and the Sustainability of Economics', *Ecological Economics*, vol. 4, nr. 2, pp. 93-117.

Klabbers, J., P. Vellinga, R. Swart, A. van Ulden and R. Janssen (1994) *Policy Options Addressing the Greenhouse Effect*, National Research Programme on Global Air Pollution and Climate Change, Bilthoven.

Langeweg, F. (1988) *Zorgen voor Morgen*, Nationale Milieuverkenning 1985-2010, RIVM, Samson H.D. Tjeenk Willink, Alphen a/d Rijn.

Leidelmeyer, K., F. van Wijk and F. Buys (1993) *Rondom Mobiliteit, Toekomst-verkenningen van de Mobiliteit Omgevingsverkenning*, Projectbureau Integrale Verkeers- en Vervoerstudies, The Hague.

Lever, J.H. (1993) Technical Assessment of Maglev System Concepts, *Proceedings of the 13th International Conference on Magnetically Levitated Systems and Linear Drives*, May 19-21, Argonne Ill.

Levy-Leboyer, C. and Y. Duron (1991) Global Change: 'New Challenges for Psychology', *International Journal of Psychology*, vol. 26, pp. 575-583.

Liebrand, W.B.G. and P.A.M. van Lange (1989) *Als Het Mij Maar Niets Kost; de Psychologie van Sociale Dilemma's*, Swets & Zeitlinger, Amsterdam.

Lovins, A.B. and J.W. Barnett (1993) 'Advanced Ultralight Hybrid-Electric Vehicles', in *Proceedings of the 26th International Symposium on Automotive Technology and Automation*, Dedicated Conference on Electric, Hybrid and Alternative Fuel Vehicles: Progress in Technology and Infrastructure, September 13-17, Aachen.

Luger, M. (1991) Local Factors Influencing Innovation Activities in Public R&D Organisations, paper prepared for the 31st Regional Science Association (RSA) European Congress, Lisbon.

Makkink, D.T (1994) 'Automatische Vervoerssystemen voor Groepen en Individuen', *Verkeerskunde*, vol. 45, nr. 1, pp. 30-34.

Marchetti, C. (1987) *Transport in Europe: The Last 50 Years and the Next 20*, IIASA, Laxenburg.

Marchetti, C. (1994) 'Anthropological Invariants in Travel Behavior' *Technological Forecasting and Social Change*, vol. 47, nr. 1, pp. 75-88.

Marcou, G. (1993) 'Public and Private Sectors in the Delivery of Public Infrastructure: the Case of the Channel Tunnel from an International Perspective', *Environment and Planning C* , vol. 11, pp. 1-18.

Martens, A.J.M. and P.G.M. Boonekamp (1990) *Elektrische Auto en Elektriciteitsvoorziening*, ECN-I-90-036, Energy Centre of the Netherlands (ECN) Policy Studies, Petten.

Martin, D.J. and L.A. Michaelis (1992) *Research and Technology Strategy to Help Overcome the Environmental Problems in Relation to Transport* (Sast Project Nr. 3), Global Pollution Study, EUR-14713-EN, Commission of the EC, Luxembourg.

Masada, E. (1993) Development of Maglev Transportation in Japan: Present State and Future Prospects, *Proceedings of the 13th International Conference on Magnetically Levitated Systems and Linear Drives*, May 19-21, pp. 1-6, Argonne Ill.

Masser, I., O. Svidén and M. Wegener (1992) *The Geography of Europe's Futures*, Belhaven, London.

Masser, I., O. Svidén and M. Wegener (1993) 'Europe 2020: Long-term Scenarios of Transport and Communications in Europe', pp. 19-41, in Nijkamp, P. (ed.), *Europe on the Move*, Avebury, Aldershot.

McClintock, M.C. (1993) The Environmental Impact of MAGLEV: A Comparative Study, *Proceedings of the 13th International Conference on Magnetically Levitated Systems and Linear Drives*, May 19-21, Argonne Ill., pp. 452-456.

McKay, C. (1989) *Möglichkeiten der Privatwirtschaftlichen Finanzierung von Verkehrsinfrastruktur-Investionen in der EG*, Beitrage aus dem Institut für Verkehrswissenschaft, nr. 19, Universität Münster.

McKenzie, J.J. (1994) 'Sustainable Vehicles for the 21st Century: Beginning the Transition to Electric and Hydrogen Vehicles', *IATSS Research*, vol. 18, nr. 2, pp. 64-70.

Messner, S. and N. Nakićenović (1993) *A Comparative Assessment of Different Options to Reduce CO_2 Emissions*, IIASA-report RR-93-15, Laxenburg.

Michaelis, L. (1995) 'The Abatement of Air Pollution from Motor Vehicles; the Role of Alternative Fuels', *Journal of Transport Economics and Policy*, vol. 29, nr. 1, pp. 71-84.

Michon, J.A. (1980) *Beïnvloeding van de Mobiliteit; Nieuwe Impulsen voor een Sturend Beleid*, Serie Verkeerskunde en Verkeerstechniek, nr. 15, ANWB, The Hague.

Miller, L. and M. Wackers (1993) 'Transrapid Maglev System: Technical Readiness and Corridors of Application', *Journal of Advanced Transportation*, vol. 27, nr. 1, pp. 49-64.

Ministry of Housing, Physical Planning and The Environment (VROM) (1991) *Fourth Report Extra* (VINEX II), The Hague.

Miyata, S., K. Matsuda, and M. Minemoto (1993) Status Quo of Development of Superconducting Maglev System in Japan, Proceedings of the 13th International Conference on Magnetically Levitated Systems and Linear Drives, May 19-21, pp. 7-9, Argonne Ill.

Monitor/SAST-activity (1994) The Environmental Problems in Relation to Transport, nr. 07/94, Commission of the EC, Brussels/Luxembourg.

Nakićenović, N. (1991) 'Diffusion of Pervasive Systems: A Case of Transport Infrastructures' in Nakićenović, N., and A. Grübler (eds.) (1991) *Diffusion of Technologies and Social Behaviour*, Springer-Verlag, Berlin, pp. 483-510.

Newman, P.W.G. and J.F. Kenworthy (1989) 'Gasoline Consumption and Cities; a Comparison of U.S. Cities with a Global Survey', *Journal of the American Planning Association*, vol. 55, pp. 24-37.

Nijkamp, P. (1993) 'Towards a Network of Regions: The United States of Europe', *European Planning Studies*, vol. 1, nr. 2, pp. 149-168.

Nijkamp, P. (1994) 'Roads toward Environmentally Sustainable Transport', *Transportation Research*, vol. 28A, nr. 4, pp. 261-271.

Nijkamp, P. and E. Blaas (1994) *Impact Assessment and Evaluation in Transportation Planning*, Kluwer, Dordrecht.

Nijkamp, P. and G.C. Pepping (1995) The Relevance and Use of Information and Telecommunication Networks as Strategic Tools in the Transport Sector: A Dutch Case Study, *Research Memorandum*, nr. 1995-5, FEWEC, Free University, Amsterdam.

Nijkamp, P. and S.A. Rienstra (1995) 'Private Sector Involvement in Financing and Operating Transport Infrastructure', *Annals of Regional Science*, vol. 29, nr. 2, pp. 221-235.

Nijkamp, P. and S.A. Rienstra (1997) 'The Public Private Nexus in Financing Infrastructure Investment', in Button, K., P. Nijkamp en H. Priemus (eds.), *European Transport Networks*, Edward Elgar, London (forthcoming).

Nijkamp, P. and J.M. Vleugel (1993) 'Success Factors for High Speed Rail Networks in Europe', *International Journal of Transport Economics*, vol. 20, nr. 3, pp. 255-270.

Nijkamp, P., P. Rietveld and H. Voogd (1990) *Multicriteria Evaluation in Physical Planning*, North-Holland, Amsterdam.

Nijkamp, P., J.M. Vleugel, R. Maggi and I. Masser (1994) *Missing Transport Networks in Europe*, Avebury, Aldershot.

Nijkamp, P., R. Camagni and R. Capello (1996a) 'Sustainable City Policy', pp. 35-55, in G.G. van der Meulen and P.A. Erkelens (eds.), *Urban Habitat*, Technical University, Eindhoven.

Nijkamp, P., G.C. Pepping and D. Banister (1996b) *Telematics and Transport Behaviour*, Springer-Verlag, Berlin.

Nijkamp, P., S.A. Rienstra and J.M. Vleugel (1997) 'Long Run Scenarios for Surface Transport', *Technology in Pollution*, vol. 7, nr. 3, pp. 305-326.

NRC (1994) Zwevend van Hamburg naar Berlijn, March 3.

NS (1988) *Rail 21; Sporen naar een Nieuwe Eeuw*, Netherlands Railways, Utrecht.

OECD (1987a) *Dynamic Traffic Management in Urban and Suburban Road Systems*, Organization for Economic Cooperation and Development, Paris.

OECD (1987b) *Toll Financing and Private Sector Involvement in Road Infrastructure Development*, Organization for Economic Cooperation and Development, Paris.

OECD (1988a) *Route Guidance and In-Car Communication Systems*, Organisation for Economic Cooperation and Development, Paris.

OECD (1988b) *Transport and the Environment*, Organisation for Economic Cooperation and Development, Paris.

OECD (1992) *Intelligent Vehicle Highway Systems; Review of Field Trials*, Organisation for Economic Cooperation and Development, Paris.

OECD (1993a), *OECD Environmental Data; Compendium 1993*, Organisation for Economic Cooperation and Development, Paris.

OECD (1993b) *Cars and Climate Change*, Organisation for Economic Cooperation and Development, Paris.

Okken, P.A. (1991) *De Betekenis van Nieuwe Energietechnologie voor CO_2-Reductie*, ECN-Rx-91-117, Energy Centre of the Netherlands (ECN) Policy Studies, Petten.

Okken, P.A. (1992) *Waterstof Energietoepassingen*, ECN-C--92-065, Energy Centre of the Netherlands (ECN) Policy Studies, Petten.

Okken, P.A., J.R. Ybema, D. Gerbers, T. Kram and P. Lako (1991) *The Challenge of Drastic CO_2 Reduction*, ECN-C--91-009, Energy Centre of the Netherlands (ECN) Policy Studies, Petten.

Okken, P.A., T. Kram, P. Lako, J.R. Ybema, J. van Doorn and D. Gerbers (1993a) *Drastische CO_2-reductie: Hoe is het Mogelijk*, ECN-C--92-066, Energy Centre of the Netherlands (ECN) Policy Studies, Petten.

Okken, P.A., T. Kram, P. Lako, J.R. Ybema and D. Gerbers (1993b) *Kansrijke Energietechnieken: Vier Blauwdrukken voor een Energiehuishouding met Minimale Emissies op Langere Termijn*, ECN-C--93-036, Energy Centre of the Netherlands (ECN) Policy Studies, Petten.

Olson, M. (1965) *The Logic of Collective Actions: Public Goods and the Theory of Groups*, Harvard University Press, Cambridge.

Onnen, M.E.K., and C.W.F. van Knippenberg (1989) *Vervoermiddelkeuze: Wat Beweegt de Mens*, VK 89-14, Verkeerskundig Studie Centrum (VSC), Haren.

Oostvoorn, F. van and W. Pellekaan (1995) *Energy Scenarios for a Changing Europe*, ECN, Energy Centre of the Netherlands (ECN) Policy Studies, Petten (confidential).

Opschoor, J.B. (ed.) (1992) *Environment, Economy and Sustainable Development*, Wolters-Noordhoff, Grcningen.

Opschoor, J.B. (1994) Institutional Change and Development towards Sustainability, *Research Memorandum*, nr. 1994-42, FEWEC, Free University, Amsterdam.

Opschoor, J.B. and R. Weterings (1994) 'Environmental Utilisation Space: An Introduction', *Milieu*, vol. 9 (1994) pp. 198-206.

Ostrom, E. (1990) *Governing the Commons*, Cambridge University Press, Cambridge.

Owens, S. (1992) 'Energy, Environmental Sustainability and Land Use Planning', pp. 79-105, in Breheny, M. (ed.), *Sustainable Development and Urban Form*, Pion., London.

Owens, S. and P. Rickaby (1992) 'Settlements and Energy Revisited', *Built Environment*, vol. 18, pp. 247-252.

Perrels, A.H. and L.P. Schippers (1991) *Eerste Raming van de Marktomvang van Hogesnelheidstunneltransport*, ESI-FUA/ISDS-study, Rijkswaterstaat, Delft.

Polytechnisch Weekblad 7/8 (1991) Burgerluchtvaart Geeft Gas!, July, p. 32.

Polytechnisch Weekblad 29/30 (1993) July 23, p. 7.

POSSUM (1997) *Strategic Policy Issues and Methodological Framework*, Background Document of Deliverable 1 of the DG-VII Possum (Policy Scenarios for Sustainable Mobility) project, Free University, Amsterdam.

Pronk, M., Klooster, J. and B. van Wee (1992) 'Het Personenautopark in 2015 Volgens FACTS 2.0 en de Gevolgen voor de Overheidsdoelstellingen', pp. 1017-1035, in Blok, P.M. (ed.), *Bundeling van Bijdragen aan het Colloquium Vervoersplanologisch Speurwerk*, CVS, Rotterdam.

Raschbichler, H.G., L. Miller and M. Wackers (1993) Analysis of Prospective TRANSRAPID Applications, *Proceedings of the 13th International Conference on Magnetically Levitated Systems and Linear Drives*, May 19-21, pp. 29-34, Argonne Ill.

Rico, A. (1993) *The Case of Mexico, Perspectives for the Possibilities of Developing Countries Giving Technological Leaps in the Field of Transportation*, CAETS-Convocation 1993, paper nr. 3, Zurich.

Rienstra, S.A. and P. Nijkamp (1997a), 'From the Expected to the Desired Future of Passenger Transport; A Stakeholder Approach' in Requier-Desjardins, D., C. Spash and J. van der Straaten, *Environmental Policies and Societal Aims*, Kluwer, Dordrecht (forthcoming).

Rienstra, S.A and P. Nijkamp (1997b) 'Lessons from Private Financing of Transport Infrastructure: Dutch Infrastructure in the 19th Century and European Projects in the 20th Century', *Revue Economique*, vol. 48, nr. 2, pp. 231-246.

Rienstra, S.A. and P. Rietveld (1996) 'Speed Behaviour of Car Drivers; A

Statistical Analysis of Acceptance of Speed Controls in the Netherlands', *Transportation Research D*, vol. 1, nr. 2, pp. 97-110.

Rienstra, S.A., J.M. Vleugel and P. Nijkamp (1996) 'Options for Sustainable Passenger Transport; An Assessment of Policy Choices', *Transportation Planning and Technology*, vol. 19, pp. 221-233.

Rietveld (1997) 'Political Economy Issues of Environmental Friendly Transport Policies' *International Journal of Environment and Technology*, vol. 7, nr. 3, pp. 398-416.

Rietveld, P. and P. Nijkamp (1993) 'Transport and Regional Development', pp. 131-151, in Polak, J. and A. Heertje (eds.), *European Transport Economics*, Blackwell, Oxford.

Robinson, J.B. (1990) Futures Under Glass; A Recipe for People Who Hate to Predict, *Futures*, October.

Ronis, D.L., J.F. Yates and J.P. Kirscht (1989) 'Attitudes, Decisions and Habits as Determinants of Repeated Behavior' in Pratkanis, A.R., S.J. Breckler and A.G. Greenwald (eds.), *Attitude Structure and Function*, Lawrence Erlbaum Associates, Hillsdale.

Rooijers, A.J. (1992) 'Maatschappelijke Acceptatie van Mobiliteitsbeleid' in Blok, P.M. (ed.), *Colloquium Vervoersplanologisch Speurwerk*, Rotterdam.

Rooijers, A.J. (1992a) *Meningen en Motieven ten Aanzien van Autogebruik voor Woon/Werk verkeer*, VK 91-14, Verkeerskundig Studiecentrum, Haren.

Rooijers, A.J. and E.M. Steg (1991) *De Rol van Gewoontegedrag bij Vervoermiddelkeuze*, Verkeerskundig Studiecentrum, Haren.

RPD (1991) *Perspectives in Europe; Exploring Options for a European Spatial Policy for North-Western Europe*, Department of Housing, Physical Planning and Environment, The Hague.

Rühl, A. (1991) Financial Relations between European Railways for their International Services, *Transportation Research*, vol. 25A, nr. 4, pp. 203-207.

Rutten, B., M. Elderman, B. van der Haspel and A. Bleijenberg (1990) *Reductie van de CO_2-uitstoot van de Personenauto door Voertuigtechnische Maatregelen Beschikbaar op Korte Termijn*, Centrum voor Energiebesparing en Schone Technologie (CE), Delft.

Salomon, I., P.H.L. Bovy and J.P. Orfeuil (eds.) (1993) *A Billion Trips a Day*, Kluwer, Dordrecht.

Samaris, Z. and K.H. Zierock (1992) *Assessment of the Effect in EC Member States of the Implementation of Policy Measures for CO_2 Reduction in the Transport Sector*, report for DG XI-B/4, Brussels.

SBB (1985) *Bahn 2000*, Schweizerische Bundesbahnen, Bern.

Schoemaker, Th.J.H., H.C. van Evert and M.G. van den Heuvel (1988) *Trendbreukscenario Vervoer en Verkeer*, Dept. Civiele Techniek, nr. VK 5101-301, Technical University, Delft.

Schmidt, T. (1989) 'Straf en Beloning als Instrument om de Modal Split te Beïnvloeden', *Verkeerskunde*, vol. 40, nr. 7/8, pp. 332-335.
Schol, E. and R.T.M. Smokers (1993) 'De Toekomst van Electrische Auto's in Nederland', *Verkeerskunde*, vol. 44, nr. 6, pp. 16-20.
Schumpeter, J.A. (1939) *Business Cycles: A Theoretical, Historical and Statistical Analysis of the Capitalist Process*, Vol. I and II, McGraw-Hill, New York.
Schwarz, B. (1988) 'Forecasting and Scenarios', pp. 327-367, in Miser, H.J., and S. Edward (eds.), *Handbook of Systems Analysis*, Wiley, New York.
Scott, A.J. (1988) *New Industrial Spaces: Flexible Production Organization and Regional Development in North-America and Western-Europe*, Pion, London.
Shladover, S.E. (1993) 'Potential Contributions of Intelligent Vehicle/Highway Systems to Reducing Transport's Greenhouse Gas Production', *Transportation Research*, vol. 27A, nr. 3, pp. 207-216.
Short, J. (1992) Environment; Global and Local Effects, Paper ECMT Conference on Transport Growth in Question, European Conference of Ministers of Transport, Lisbon.
Smokers, R.T.M. (1993) 'Electric Vehicles in the Context of New Transportation Systems', pp. 77-84, in *Proceedings of the 26th International Symposium on Automotive Technology and Automation*, Dedicated Conference on Electric, Hybrid and Alternative Fuel Vehicles: Progress in Technology & Infrastructure, 13-17 September, Aachen.
Smokers, R.T.M. and O. van Hilten (1992) *Elektrische Auto's in het Perspectief van de Milieu- en Energieproblematiek*, ECN-RX--92-073, Energy Centre of the Netherlands (ECN) Policy Studies, Petten.
Soekkha, H.M., P.H.L. Bovy, P. Drewe and G.R.M. Jansen (eds.) (1990) *Telematics - Transportation and Spatial Development*, VSP, The Hague.
Spee, B.M. and H. Wittenberg (1978) Luchtvaart en Energiegebruik, *Tijdschrift voor Vervoerswetenschap*, vol. 14, nr. 3, pp. 236-256.
Stam, W.J. (1991) *Technische Ontwikkeling en Mobiliteit*, ITL-series nr. 4, Infrastructuur, Transport en Logistiek, Delft University Press, Delft.
Steen, P. (1994) Scenario Methods; Backcasting versus Forecasting, paper presented at the 2nd network seminar on Sustainable Mobility Scenarios, 24-25 May, Aalborg.
Stokes, G. and S. Hallett (1992) 'The Role of Advertising and the Car' *Transport Reviews*, vol. 12, nr. 2, pp. 171-183.
Stuurgroep Ondergrondse Vervoersinfrastructuur (1993) *Ondergronds Overwegen*, SDU, State Printing Office, The Hague.
Svidén, O. (1990) A Scenario on the Evolving Road Service Informatics in Europe, *Transportation*, vol. 17, pp. 231-237.
Svidén, O. (1989) *Scenarios, on Expert Generated Scenarios for Long Range Infrastructure Planning of Transportation and Energy Systems*, Linköping

Studies in Management and Economics, Dissertations nr. 19, University Linköping.

Swahn, H., U. Björketun and J. Eriksson (1994) Reduced CO_2 Emissions through Higher Petrol Prices, *Nordic Road and Transport Research*, nr. 3, pp. 32-33.

Taniguchi, M. (1995a) The Japanese Shinkansen, *Built Environment*, (forthcoming).

Taniguchi, M. (1995b) 'The Japanese Magnetic Levitation Trains', *Built Environment* (forthcoming).

Taylor, B. (1990) 'Strategic Planning - Which Style Do You Need?', pp. 33-51, in Hahn, D. and B. Taylor (eds.), *Strategische Unternehmungsplanung - Strategische Unternehmungsführung*, Physica Verlag, Hamburg.

Technieuws 29, nr. 1 (1991) 'Train à Grande Vitesse- De "Concorde op Rails"', Ministry of Economic Affairs, August.

Thord, R. (ed.) (1993) *The Future of Transportation and Communication*, Springer-Verlag, Berlin.

Tietenberg, T.H. (1992) *Environmental and Natural Resource Economics*, Harper Collins, New York.

TRB (1991) *In Pursuit of Speed, New Options for Intercity Passenger Transport*, Transportation Research Board, Washington D.C.

Tweede Kamer (1990) *Tweede Structuurschema Verkeer en Vervoer; Deel d: Regeringsbeslissing*, meeting year 1989/1990, 20 922, nr. 15-16, SDU, State Printing Office, The Hague.

Tweede Kamer (1991) *Nederlands Deel Hogesnelheidsspoorverbinding Amsterdam-Brussel-Parijs*, 22 026, SDU, State Printing Office, The Hague.

Verhoef, E.T. (1994) 'External and Social Costs of Road Transport', *Transportation Research*, vol. 28A, nr. 4, pp. 273-287.

Verhoef, E.T. (1996a), *The Economics of Regulating Transport*, Elgar, Cheltenham.

Verhoef, E.T. (1996b) *The Regulation of Road Transport Externalities*, Ph.D. thesis, Free University, Thesis Publishers, Amsterdam.

Verhoef, E.T., P. Nijkamp and P. Rietveld (1997) Tradeable Permits: Their Potential in the Regulation of Road Transport Externalities, *Environment and Planning B*, 1997.

Versuchs- und Planungsgesellschaft für Magnetbahnsysteme mbH (1985) *Zwischen Fahren und Fliegen, TRANSRAPID im Verbund mit Auto, Eisenbahn und Flugzeug.*

Vickerman, R.W. (1991) 'Other Regions' Infrastructure in a Region's Development', pp. 61-74, in Vickerman, R.W. (ed.), *Infrastructure and Regional Development*, Pion, London.

Victor, D.G. (1990) 'Liquid Hydrogen Aircraft and the Greenhouse Effect', *International Journal on Hydrogen Energy*, vol. 15, nr. 5, pp. 357-367.

Vlek, C. and J.A. Michon (1992) 'Why We Should and How We Could Decrease the Use of Motor Vehicles in the Near Future', *IATTS-Research*, vol. 15, nr. 2, pp. 82-93.

Vleugel, J.M. (1995) *Milieugebruiksruimte voor Duurzaam Verkeer en Vervoer; een Analyse van de Toepasbaarheid voor Beleid*, ITL nr. 21, Infrastructuur, Transport en Logistiek, Delft University Press, Delft.

Voogd, H. (1983) *Multicriteria Evaluation for Urban and Regional Planning*, Pion, London.

Vranich, J. (1992) High Speed Hopes Soar, *Railway Age*, May, pp. 22-32.

WCED (World Commission on Environment and Development) (1987) *Our Common Future*, (Brundtland Report), Oxford University Press, Oxford.

Wegener, M. (1995) 'Reduction of CO_2 Emissions of Transport by Reorganisation of Urban Activities' in Hayashi, Y. and J.R. Roy (eds.), Transport, Land Use and the Environment, Kluwer, Dordrecht.

Wenke, R. (1992) 'Verkehrinfrastrukturpolitik in der EG', *Zeitschrift für Verkehrswissenschaft*, vol. 63, nr. 3, pp. 180-188.

Whitelegg, J. (1993) *Transport for a Sustainable Future*, Belhaven Press, London.

Wiescholek, U. and W.J. Mayer (1992) High-Speed Magnetic Levitation Train Transrapid, Planning of the Final Development until 1995 and Prospects of Utilization in the Federal Republic of Germany, *Selected Proceedings of the 6th World Conference on Transport Research*, vol. II, Lyon, pp. 3123-3134.

Wissman, M. (1994) 'German Transport Policy after Reunification' *Transportation Research*, vol. 28A, nr. 6, pp. 453-458.

Wit, J. de and H. van Gent (1996) *Economie en Transport*, Lemma, Utrecht.

Witsen, M. van, T.J.H. Schoemaker and P.B.L. Wiggenraad (1987) *Openbaar Vervoervoorzieningen*, Traffic Dept., Technical University, Delft.

WMO (1988) World Conference on the Changing Atmosphere, Implications for Global Security, *4s Proceedings Toronto 1988*, WMO report nr. 710, Geneva.

Wood, C., S. Watson and D. Banister (1994) The Relationship between Energy Use in Transport and Urban Form, Paper for Universities' Transport Studies Group 1994 Conference, 5-7 January, University of Leeds.

Ybema, J.R. and P.A. Okken (1993) *Full Fuel Chains and the Basket of Greenhouse Gases*, nr. ECN-C--93-050, Energy Centre of the Netherlands (ECN) Policy Studies, Petten.

Zeevenhooven, N.H.C.E. van (1993) Technische Eisen en Criteria te Stellen aan Vervoersystemen; in het Bijzonder aan Geleide Vervoertechnieken, *Nieuwe Vervoerssystemen en Technieken*, Stichting Postacademisch Onderwijs in de Vervoerswetenschappen en de Verkeerskunde, Delft.

Zwier, R., F. Hiemstra, P. Nijkamp and K. van Montfort (1994) Connectivity and Isolation in Transport Networks: A Policy Scenario Experiment for the

Greek Island Economy, *Research Memorandum*, nr. 1994-55, FEWEC, Free University, Amsterdam.

INDEX